Radical Theologies

Radical Theologies is a call for transformational theologies that break out of traditional locations and approaches. The rhizomic ethos of radical theologies enables the series to engage with an ever-expanding radical expression and critique of theologies that have entered or seek to enter the public sphere, arising from the continued turn to religion and especially radical theology in politics, social sciences, philosophy, theory, cultural, and literary studies. The post-theistic theology both driving and arising from these intersections is the focus of this series.

Religion, Politics, and the Earth: The New Materialism
By Clayton Crockett and Jeffrey W. Robbins

The Apocalyptic Trinity
By Thomas J. J. Altizer

Foucault/Paul: Subjects of Power
By Sophie Fuggle

A Non-Philosophical Theory of Nature: Ecologies of Thought
By Anthony Paul Smith

On Philosophy as a Spiritual Exercise: A Symposium
Edited by Philip Goodchild

The Counter-Narratives of Radical Theology and Popular Music: Songs of Fear and Trembling
Edited by Mike Grimshaw

Theology after the Birth of God: Atheist Conceptions in Cognition and Culture
By F. LeRon Shults

Theopoetics of the Word: A New Beginning of Word and World
By Gabriel Vahanian; Foreword by Noëlle Vahanian

Economics in Spirit and Truth: A Moral Philosophy of Finance
By Nimi Wariboko

Retrieving the Radical Tillich
Edited by Russell Manning

Kierkegaard and the Refusal of Transcendence
By Steven Shakespeare

Religious Experience and New Materialism: Movement Matters
Edited by Joerg Rieger and Edward Waggoner

Religious Experience and New Materialism

Movement Matters

Edited by

Joerg Rieger and Edward Waggoner

palgrave
macmillan

RELIGIOUS EXPERIENCE AND NEW MATERIALISM

First published 2016 by
PALGRAVE MACMILLAN

Palgrave Macmillan in the UK is an imprint of Macmillan Publishers Limited, registered in England, company number 785998, of Houndmills, Basingstoke, Hampshire, RG21 6XS.

Palgrave Macmillan in the US is a division of Nature America, Inc., One New York Plaza, Suite 4500, New York, NY 10004-1562.

Palgrave Macmillan is the global academic imprint of the above companies and has companies and representatives throughout the world.

Hardback ISBN: 978–1–137–56843–4
E-PUB ISBN: 978–1–137–56845–8
E-PDF ISBN: 978–1–137–56844–1
DOI: 10.1057/9781137568441

Distribution in the UK, Europe and the rest of the world is by Palgrave Macmillan®, a division of Macmillan Publishers Limited, registered in England, company number 785998, of Houndmills, Basingstoke, Hampshire RG21 6XS.

Library of Congress Cataloging-in-Publication Data is available from the Library of Congress.

A catalogue record for the book is available from the British Library.

Contents

Series Preface

Radical Theologies encompasses the intersections of constructive theology, secular theology, death of god theologies, political theologies, continental thought, and contemporary culture.

For too long, Radical theology has been wandering in the wilderness, while other forms of theological discourse have been pontificating to increasingly smaller audiences. However, there has been a cross-disciplinary rediscovery and turn to radical theologies as locations from which to engage with the multiplicities of twenty-first century society, wherein the radical voice is also increasingly a theologically-engaged voice with the recovery and rediscovery of radical theology as that which speaks the critique of "truth to power."

Radical Theologies reintroduces radical theological discourse into the public eye, debate, and discussion by covering the engagement of radical theology with culture, society, literature, politics, philosophy, and the discipline of religion.

Providing an outlet for those writing and thinking at the intersections of these areas with radical theology, *Radical Theologies* expresses an inter-disciplinary engagement and approach that was being undertaken without a current series to situate itself within. This series, the first dedicated to radical theology, is also dedicated to redefining the very terms of theology as a concept and practice.

Just as Rhizomic thought engages with multiplicities and counters dualistic and prescriptive approaches, this series offers a timely outlet for an expanding field of "breakout" radical theologies that seek to redefine the very terms of theology. This includes work on and about the so-labeled death of god theologies, theologians who emerged in the 1960s, and those who follow in their wake. Other radical theologies emerge from what can be termed underground theologies and also a/theological foundations. All share the aim and expression of breaking out of walls previously ideologically invisible.

Acknowledgments

The co-editors would like to thank SMU Graduate Students Ben Robinson (Religion and Culture) and Timothy McGee (Systematic Theology) for their support in editing and preparing the manuscript for this book. Special thanks go to Timothy McGee for compiling the index.

Introduction

New Materialism and the Study of Religion

Tamsin Jones

Introduction

The application of new materialism—its methods, assumptions, and aims—to the study of religion is part of a larger "material turn" in the field. The precise elements of this turn might vary according to who is detailing them; however, all would agree that it involves, at least in large part, a rejection of the interiority, ideality, and emphasis on transcendence that long held sway in considerations of religion, in favor of exteriority, materiality, and immanence. Within this broader material turn, one of the defining elements of *new* materialism is the further rejection of an anthropocentrism central to much modern religious and theological thought.[1] It is not much of a stretch to say that new materialism and religion, at first glance, seem to be a rather uncomfortable pairing. It is not surprising, then, that, despite some call for a new materialist mandate for the study of religion,[2] there has not as yet been a proliferation of new materialist analyses of religion. Nonetheless, as is often the case, sites of tension yield creative and intriguing work.

This volume seeks not only to add to the growing collection of studies of religion, which take into account the insights of new materialism, but it also critically engages earlier attempts to do so. In particular, many of the essays included herein respond (if sometimes obliquely) to the insights and challenges of the most comprehensive experiment of reading religion and theology through a new materialist lens, to date: Clayton Crocket and Jeffrey Robbins's *Religion, Politics and the Earth: The New Materialism*.[3] New materialism is about *movement*—bodily, political, organizational—as much as discrete material realities. The essays collected here expand upon the site of materiality by focusing in particular on this dynamic movement of the material reality of religion. Given that movement contains tensions, the volume as a whole consciously allows some of the disagreements within new materialist interpretations of religion to be present, to challenge one another, and to remain in disagreement. By way of introduction, I will survey the main arcs of the disparate and elusive collection of discourses labeled "new materialist." To do so, I rely primarily on two recent introductions to new materialism,[4] before outlining some of the issues pertaining to the interaction of new materialism and the study of religion more specifically.

Mapping New Materialism

New materialism includes voices from feminist theory, political philosophy, cultural and race theory, anthropology, and sociology. It draws from an even broader net, however, including concepts and insights from physics, biology (bioethics, biomedicine, and biotechnology), media theory, aesthetics, and ecological and environmental sciences. As a result of being radically interdisciplinary as well as "in many ways a wave approaching its crest,"[5] new materialism is not a clearly defined single movement that can be easily summarized. For this reason, in order to draw their "cartographies" of the shifting terrain of new materialism, Dolphijn and van der Tuin identify "transversality"—the intersecting of multiple discourses—as one of the primary features of new materialism. Following Dolphijn and van der Tuin, rather than trying to provide a single overarching definition of new materialism, this introduction shall instead attempt to map out not only the sources of its emergence, but also the primary lines of inquiry that new materialists follow, and the main ideas developed therein.

Sources of New Materialism

New materialism emerges out of a critical shift in certain circles within feminist theory and in late twentieth century continental philosophy. First, theorists such as Elizabeth Grosz, Donna Haraway, Rosi Braidotti, and Karen Barad diversely make arguments that seek to root feminist theories in material contexts and realities without a concurrent return to essentialism.[6] Second, the speculative turn in philosophy seeks to elucidate a metaphysical realism without resorting to the transcendent moves made within the "theological turn" in late twentieth century Continental—especially French—philosophy. The term "speculative realism"—though there are some problems with the appellation[7]—serves as a marker for a loose collection of French philosophers who have given renewed attention to a consideration of metaphysics in a realist and materialist tenor. In such a grouping, one would find such thinkers as Quentin Meillassoux, Bruno Latour, and Catherine Malabou.[8] One of the claims held in common is a rejection of the concept of "correlationism," in which there is a structure of relation between humans and world, thought and being. Speculative realism makes a consequent assumption about the "autonomy of reality"—that is, it considers "reality as it is absolutely independent of humans considering it, where it is possible, to consider what the world was like—or will be like—without human beings."[9] It is clear how these two disparate intellectual shifts give way to a common aim as it is articulated in the emerging new materialism: namely, a shift to material realities and away from linguistic and discursive ones.

Beyond these broad seismic shifts in feminist theory and continental philosophy, Coole and Frost's excellent introductory text identifies a number of more specific movements in response to which the new materialist voice emerges. While being a little late on the uptake, social theorists have caught on to some of the implications of the revolutions in the natural sciences, which occurred much earlier in the twentieth century. Thus, new materialism emerges, in part, out of the awareness that matter is not in fact dormant, passive, inert stuff that gets acted upon. From both physics and biology, matter is recognized to be both "considerably more elusive . . . and complex."[10] Thus, rather than continuing in the classical (Cartesian-Newtonian) way of treating matter as a "brute 'thereness'" in which there is a linear relationship between cause and effect—a relationship which determines the human knowing subject as the supreme cause and master over the nature it effects[11]—new materialism not only seeks to avoid, first and foremost, any dualism between spirit/mind and matter, but also recognizes that materiality is always more than "mere matter."[12] Further, the constant movement of matter unsettles the stable grounds we assume in daily life and thought.[13] Rather than the assumption of a dualism between spirit and matter, new materialists postulate monism.

The second example of a non-cognate discipline that nonetheless undergirds new materialism is found in the rapidly developing life sciences (bioethics, biomedicine, and biotechnology), in which the porosity and dynamism of organisms likewise give the lie to the simplistic notion of matter as a discrete entity with clearly defined borders.[14] Furthermore, these biosciences indicate and produce dramatic changes in our understanding of what human life entails. There is no aspect of our lives that is unaffected by developments in biotechnology, whether it is the genetically-modified food we eat, or fuel we require, or any number of medical and digital prostheses we use: pacemakers to regulate the beating of our hearts, Skype to communicate with family on the other side of an ocean, GPS to locate us in space and direct us away from traffic jams, Twitter feeds (to either foment revolution or follow the minutiae of celebrity existence), psychotropic drugs to modify our moods and emotional reactions, and so on. These biotechnical advances have made the distinction between "nature" and "culture" at least arbitrary, if not nonsensical.

Common Themes in New Materialist Thought

New materialism remains firmly located at the tail end of postmodernity. One indication of this is the fact that it neither seeks to return to any reactionary traditionalism, nor to embrace a scientific positivism, but remains explicitly open-ended and indeterminate. Indeed these qualities are repeated frequently throughout both introductions referenced here.[15] Nevertheless, one Enlightenment hangover is retained, indeed expanded: namely, the value of a particular form of freedom, as the term is used in modernity. The reason that indeterminacy is so valued is that it indicates the freedom, not just of thinking subjects, but of all material phenomena. For this reason, any theory of materiality will be radically pluralistic. There is no one single unified doctrine of new materialism—a fact acknowledged structurally by the three main English language introductions to new materialism, which have multiple authors.[16]

Second, and relatedly, new materialism posits not simply the significance, but also the *agency*, of the material world, rather than continuing with the fable of a subject who acts upon dormant and inert matter. Instead, assuming the "force of materiality" new materialists will consider the "liveliness of living matter"[17] or, as Jane Bennett puts it, the "agency of matter." In her contribution to Coole and Frost's volume, Bennett, who is most identified with this claim, is careful to clarify that agency in this context does not entail intentionality: as she writes, she is interested in "historical lines of thought in which materiality is figured not as inert or even passively resistant but as active and energetic, albeit *not as purposive in any strong sense*."[18] Active and energetic, yet nonpurposive, non-intentional agency is a way of describing the inevitable impact of matter on each and every experience. For instance, in her most well-known work *Vibrant Matter*,[19] when she examined the agency of electricity in the historical event of a blackout, Bennett considers electricity as a bona fide agent on its own terms rather than as one of many "instrumentalities, techniques of power, recalcitrant objects, or social constructs."[20]

What is significant about a discussion of the agency of matter is its displacement of the human subject as the only agent, the sovereign lord over his domain of knowledge and action. Crockett and Robbins make the same move when they posit the "earth as subject."[21] Further, the proclamation of such a "flat ontology" as we find in Crockett and Reader's contribution to this volume is part of a critical demand for the human to relinquish its role of master of the universe, and instead recognize "the human as always already part of the non-human in its broadest sense," thus challenging "any notion of human exceptionalism." For this reason, new materialism situates itself within a "posthuman" world. Posthumanism not only refers to the displacement of the human subject as the center of reality, the sovereign master of the universe; it also challenges the notion that there is any longer such a discrete entity in nature as the human.

The vast and intricate interconnectedness of our lives, and the impact on the planet and the organisms (human, animal, and plant) living in it, is another fact of the reality we inhabit and to which must respond intellectually.[22] The tight interaction between the environment, biology, bodies, the economy, and politics means there is no neutral action. Every decision we make, or action we take, has far-flung repercussions. As Crockett and Reader make clear in their essay in this volume, one can trace the complex myriad of negative consequences—ecological, consumer health, the treatment of animals, impact on local economies, and the like—of the consumer demand for cheap meat and milk. And, of course, our reliance on petroleum as a cheap source of energy not only decides the political and economic power of certain nations; it sends us to war.[23]

The ethical thrust motivating much new materialism is striking. In the face of the "complex issues such as climate change or global capital and population flows, the bio-technological engineering of genetically modified organisms, or the saturation of our intimate and physical lives by digital, wireless, and virtual technologies,"[24] how can a serious thinker *not* think seriously of the material worlds that shape and are shaped by our thinking? For the reasons just rehearsed, it should be clear why new materialism has, from its beginning, had both a critical and a practical orientation. Theorists are not satisfied with a description of material realities; they must also explore "the way social and economic practices produce and reproduce embodied subjectivity and existential inequalities, as well as the spaces of, and possibilities for, political transformation."[25]

Crockett and Robbins are devastatingly cutting about this critical beginning point of their work. The book starts with the premise that we've reached the "limits of global capitalism"—a claim they state as a brute material fact—infinite growth cannot rely on finite natural resources.[26] The realization of this dilemma has triggered three interlocking crises—ecological, energy, and economic—which together form the foundation or starting place for any critical theory today. Thus opening their work with a quote from Heidegger ("The being of human being is used up"), Crockett's and Robbins's book begins with the end: "We are witnessing the exhaustion of contemporary culture, a devolution of consumerism, greed, mindless entertainment, and the corrupt appeal of money and military power."[27] Rather than confront the multiple scenarios of apocalyptic crises in play within our collective consciousness, too often we stick our heads in the sand with an increasingly delusional refusal to witness the decay and dissolution of our world—culturally, politically, economically, and of course, materially. The authors go on to list in the following pages any number of undeniable, and undeniably bleak, signs of this demise;[28] indeed, they are at their best in their critical diagnosis of the challenges facing our world.

This explicitly political and practical orientation of new materialism generates an interesting critique of the dominance of social constructivist theories of society. Such a "critical analysis of actual conditions of existence" involves new materialism in a "methodological realism that is at odds with some of the more radical, and especially linguistic, forms of constructivism."[29] Indeed, one of the clear strengths of new materialism is its "critical realism." As Coole and Frost put it, "from this materialist point of view, it is ideological naïveté to believe that significant social change can be engendered solely by reconstructing subjectivities, discourses, ethics, and identities—that is, without also altering their socioeconomic conditions or tracing crucial aspects of their reproduction to the economic interests they unwittingly serve."[30] In this sense, new materialists will claim that the constructivist position that sought to unseat the subject from sovereignty and neutral autonomy in fact remains an egocentric or subject-centered discourse.[31] New materialists by no means deny that our realities are socially constructed. It is just that the construction is not the whole story. Our realities are not merely cultural products; they are also materially given and "real," by which I mean stubbornly agential. Reality is both constituted by, and resistant to, our constructions.

In this way, and merging with the frustration that gave rise to a "speculative realism" within philosophy, new materialism expresses a certain fatigue and dissatisfaction with the inherent limits of theory dominated by the cultural or linguistic turn. As Coole and Frost put it, new materialism is a response to the "sense that the radicalism of the dominant discourses which have flourished under the cultural turn is now more or less exhausted… [in particular] the dominant constructivist orientation to social analysis is inadequate for thinking about matter, materiality, or politics in ways that do justice to the contemporary context of biopolitics and global political economy."[32] Such an "allergy to the 'real'"[33] provides no basis to respond to the material realities and challenges faced today: an overpopulated planet with shrinking natural resources to support all forms of life, climate change, the energy crisis, the vastly growing disparity between the "haves" and the "have-nots" that is the sign of the collapse of late-stage capitalism, the collapse of higher education in the face of increasing corporate infiltration, the cultural wasteland saturating our media and stupefying the general population. The acuteness of some of these challenges make many of the arguments and debates of the last few decades of cultural theory seem, at best, insufficient, and more perniciously, intellectually indulgent.[34]

The final characteristic of new materialism emerges out of its Spinozian-Deleuzian roots: new materialism eschews any claim or appeal to transcendence in favor of any entirely immanent system of thought and reality. All three introductions to new materialism make this immanent domain explicit. Dolphijn and van der Tuin express this in terms of the monist orientation of new materialism, which rejects the dualisms of matter and thought.[35] Crockett's and Robbin's rejection of transcendence comes from their rereading of Hegel along the lines of Žižek and Malabou. As counter-intuitive as it seems, where Crockett and Robbins end up is in a kind of marriage of Hegel and Deleuze in which Hegel's Spirit is really "energy"; energy is the third in the dialectic, sublating both crude matter and transcendent spirit and the ideologies that accompany both.[36]

New Materialism and Religion

With the exception of this volume and Crockett's and Robbin's work, there has not yet been much explicit work on religion from a new materialist perspective. Nonetheless, one can readily discover elements of the spiritual haunting new materialist discourses. Such echoes can be understood as the attempt to provide an alternative to what Heidegger critically diagnosed as technological thinking—the kind of thinking that enables the exploitation of "things-in-the-world" through our obsession with "economic, mathematical, and quantitative calculation." With technological thinking, all beings "take the form of 'standing reserve' (*Bestand*)," or things to use.[37]

What is the best way to counter such calculative thought? Heidegger offers one possibility himself: "meditative thinking," an alternative that Crockett and Robbins gloss as meaning "to stop grasping and trying to control reality, and to let go."[38] This is a project taken up by the "new French phenomenology" following Heidegger (by thinkers such as Emmanuel Levinas, Jean-Luc Marion, Michel Henry, et al.). However, with its emphasis on an alterity capable of interrupting our cozy world of sameness, Crockett and Robbins suspect this philosophical discourse remains too "mystical and otherworldly," or, in other words, too indebted to a mode of transcendent thinking. Nonetheless, as LaMothe discusses in her essay, many new materialist thinkers themselves do not shirk from using language that gestures to the spiritual. Thus, the question of the function and status of religion in new materialist discourse remains. What would satisfy as a non-technological but still immanent or worldly kind of thinking? If such a mode of thinking religion can be articulated, this leads to another question: what is gained and lost by tethering religion to immanence?

Crockett and Robbins give the first sustained response to this question prior to this volume. They argue that one might partner with religious thinking in a way that is acceptable, even beneficial to new materialism, provided one recognizes the mode of religious thinking in this instance as specifically Feuerbachian. This return to the original religious materialism of Feuerbach could be reduced to a championing of a "religion without God" (in the mode of Dworkin and Derrida). However, such a statement requires further explanation. What was Feuerbach's original argument about religion, and how do the new materialists make use of it? As several of the essays in this volume make some reference to Feuerbach, it may be helpful to review his original argument.

Feuerbach's originality lies in the argument that the Christian idea of God is not a reflection of some divine being, but more precisely an image of Christian believers' ideals writ on a transcendent screen. This has two primary implications: first, the contention that religion is, from start to finish, a human artifact; second, that the function of religion is to externalize and objectify our ideals. Let us discuss each of these implications in turn.

If, for Feuerbach, religion is a historical human construct, then it reveals a great deal of what it means to be human:

> Such as are a man's thoughts and dispositions, such is his God; so much worth as a man has, so much and no more has his God. Conscious of God is self consciousness, knowledge of God is self-knowledge. By his God, thou knowest the man, and by the man his God; the two are identical.[39]

Religion, it turns out, isn't about God; it's about the human. As a result, the value of religion is, primarily, that it tells us about ourselves. What does this mean for the study of religion?

Crockett and Robbins state the implications of this Feuerbachian view of religion thus: "First, the point of religious analysis is decidedly not for the sake of religion; rather it is a means to an end. Religion is used to explain humanity to itself. By claiming religious knowledge as self-knowledge, the antithesis between the divine and the human is broken down."[40] Thus, in their view, all study of religion today ought not only to be Feuerbachian; it ought to be Marxist. The study of religion should attend to how religion functions ideologically in "service of a repressive economic order," or, we might add, repressive sexual order, political order, or in the service of ecological devastation, and so on.[41] With an appropriate recovery of a more accurate reading of Marx,[42] however, they argue that the recognition of religion as anthropology does not mean it is valueless. Indeed, it becomes, not only an important diagnostic instrument to take the temperature of the human spirit, it also provides a dose of hope—even if the medicine is no more than a placebo. In other words, Crockett and Robbins want to argue that there is a *functional* value to religion: "it provides the heart and spirit where otherwise there would be none."[43]

The question that they do not engage in is whether or how religion can continue to function in this way once it has been shown to be nothing more than anthropology writ on a celestial stage, without a sense of the transcendent. Crockett and Robbins argue that "the power of religion is independent of its truth,"[44] but this begs the question of whence comes its power to inspire or give hope if it is a purely human phenomenon? I will return to this question in the next section of this introduction, where I identify some of the open questions of the volume.

However, this is not the only significance of Feuerbach. The nineteenth century philosopher also completes the anthropomorphic reorientation of our understanding of religion (one begun with the subjective turn found in Schleiermacher) in a way that grounds it in materiality. In a telling sentence, Feuerbach states that "[m]an first of all sees his nature as if *out* of himself, before he finds it in himself."[45] When Feuerbach claims that God is a human projection, he is merely making objective the subjective claims Schleiermacher made. Or, as Crockett and Robbins gloss this shift (in a way that clearly indicates its significance for a materialist), "the religious subject is not only born out of the infinitude of the power of feeling but also by its objectification of that feeling: religion becomes more than feeling. It becomes ontological. It becomes material."[46] What this does, in effect, is remove the bias in scholarly analyses of religion toward interiority. It forces the recognition of religion as always materially manifest in one way or another.[47]

These two vital shifts heralded by Feuerbach—religion as anthropomorphic and religion as externalized or objectified—follow two quite different trajectories among materialist theories of the study of religion. Very briefly we might compare the treatment of religious experience in two handbooks on the study of religion, studies which bookend the last two decades: *Critical Terms for Religious Studies*[48] from 1998 and *The Cambridge Companion to Religious Studies* from 2012.[49] In the former, religious experience is notoriously disregarded by Robert H. Sharf as indicating nothing more than one's "ideological commitments, shaped by vocation, socio-economic background, politics, sectarian affiliations, etc."[50] Thus, it would seem we study religious experience only in order to lay bare all the prejudices of those who claim them, or who claim an interest in studying them. The study of religion, in Sharf's hands, serves a merely critical function by unveiling hidden biases and ideological constructions of its participants.

This approach is interestingly compared with the very different argument found in the more recent text, *The Cambridge Companion to Religious Studies,* in which Robert Orsi not only challenges the reductive analysis of religious lives through cultural explanation,[51] he upends common assumptions about the ideality and subjectivity of religious experience itself. Through a brilliantly unexpected rehabilitation of Rudolph Otto's famous, and famously derided, essay translated into English as *The Idea of the Holy*, Orsi demonstrates how the English translation masks a shift of the experience of the holy from a concrete, real presence to the ideal, subjective cognition of one. Otto's own title was simply *Das Heilige*—The Holy. Thus, the translation into the English-speaking world of religious studies scholars requires a shift from "the straightforward realism of 'the holy' to the psychological 'idea of the holy' [and] . . . a shift from the objective to the subjective," which is precisely the opposite of Otto's intention.[52] Otto assumed this experience of the holy was real, not ideal, concrete, and present—not symbolic or imagined. Orsi reiterates this point and further shows how the experience of the holy is not subjective, ideal, private, and individual; it is material and public, manifest in sacred objects, rituals, food practices, bodily disciplines, political impacts, and so on.

Which version of Feuerbachian religion is most operative in new materialism? According to Crockett and Robbins, once thinkers like Žižek clear the way back to religion by claiming that it is not the only form of "false consciousness"—that indeed, all consciousness is in some sense "false"—and once the secular hypothesis, which assumed that clear reason and science would pave the way to a peaceful prosperous world as soon as the specter of religion had been cleared away, is itself proven naïve and mistaken, then the path to a return to religion is cleared: "No longer must the materialist rendering of religion be articulated in the language of critique alone. In fact, with the new materialism religion might become a source of political empowerment and political mobilization."[53] Without explicitly saying so, what new materialism is here suggesting, is a return to Feuerbach's original stance. The recognition that "theology is anthropology" was, for Feuerbach, *not* a claim about the falseness of religion, but a claim about its ultimate power and significance.

On the other hand, some of the essays in this volume suggest two potential questions with this approach. First of all, it is not clear that this Feuerbachian optimism is a possible stance to maintain, given the closed self-reflexivity of the system. By reducing religion to anthropology and admitting that what is worshipped is nothing but a man-made God (i.e., an "idol"), does this not take away some of the force of religion to transform given realities? A second potential problem is that such a view can end up projecting an overly smug approach to the study of religion. It can appear to convey a claim about religion that functions with the masses, while the elite gnostics—the philosophers, the religious studies scholars, the theorists—confidently identify what is "really going on." Does this not risk the danger of disregarding the voices of those religious subjects whose experience of the holy founds their movements of resistance? One of the strengths of this volume is that it does not decide in a singular way upon these questions, but makes clear what the implications of the choices are.

Expansions and Critical Engagements with New Materialism

By titling the volume *Religious Experience and New Materialism: Movement Matters*, the editors signal an emphasis on the importance of analyzing not merely the dynamic energy of matter, but also the vitalism of human movement(s). For instance, *LaMothe's* chapter articulates the necessity of attending to "the bodily dimensions of religious life" in order to develop what she names an "eco-kinetic approach to the study and practice of religion," an approach that is kinetic, empathic, imaginative, participatory, and eco-evaluative. For LaMothe, unlike Crockett and Reader, such an analysis of religion need not eschew the transcendent entirely. Moreover, it requires attention to not only the *matter* of religious practices, but also of their *movement*, in order to apprehend a notion of human being "as an ongoing rhythm of bodily becoming." It is through bodily movement, she argues, that "those qualities we prize as distinctively human—empathy, compassion, cooperation, and self-consciousness—in short, the ability to love" are developed.

Likewise, *Minister* is interested in exploring the ways in which "organizing" can be analyzed as a religious practice that is intrinsic to religious life. He discusses the organizing movements of communities as a phenomenon of conversion that "changes the orientation of bodies in society." Drawing on arguments by Sara Ahmed, Jacques Rancière, as well as Crockett and Robbins, he enables us to reconsider conversion beyond an individual (and often coerced) act of will or belief, to a more bodily and communal capacity to generate social power—a power which may "empower resistance to unsustainable modes of production that exhaust life."

Crocket and Reader take a more concrete approach through introducing specific projects within "relational Christian realism"[54] that intentionally form themselves along social and ecological premises. In so doing, they draw attention to what is oft ignored within new materialist manifestos, namely, the "growing engagement between ecological concerns as a New Social Movement and faith-based entanglements." For instance, starting from an analysis of the massive industrial pork production in Brittany, France (or North Carolina) with its correlative destructive costs (from deforestation, to polluted waterways and large spills of untreated waste, to the cruel treatment of animals and laborers), Crockett and Reader draw conclusions about human greed and the assumption of exceptionalism that is "based upon a hierarchical or metaphysical superiority of the human." While this exceptionalism may have its foundation in Christian theology, a mode of Christian thinking that is rigorously realist and immanentist might, they argue, provide the best challenge to such destructive assumptions and activities that drive the cheap industrial production of food.

Along complementary lines, *Bray* acutely challenges new materialists to consider not simply productive movements, but also those of expulsion. This consideration of ourselves as waste and waste-machines, and not merely "brains," points to the fact that creativity and new life comes from that *decomposition* of bodies as much as from the *ordering* of them. She challenges her reader to find God there too—as not only the "giver and taker on of forms," but also "in the shit pile of history, on the side of the expelled."

Finally, *Rieger* lays out not only "roadblocks for" and "contributions of" new materialist approaches to the study of religious experience, he also critically and constructively pushes past some of the limitations of new materialist studies thus far. He forcefully demonstrates the inadequacy of any analysis of our material worlds that does not appropriately foreground forces of capitalism and labor. In so doing, he pinpoints one of the weaknesses of the lack of attention to the human subject, especially the collective agency of social movements that can haunt new materialism. This idea is summed up in a pithy formula—"materialism needs movements"—a formula that can be said to anchor the volume.

This volume not only critically expands upon new materialism in order to emphasize the material structures of socio-political movements of religious groups and the physiology of religious bodies, but it does so in a way that presents the tensions and disagreements among religionists taking up new materialism. In this final section, I will highlight two of the main critical questions and tensions that the reader will encounter in the essays that follow.

The first tension has to do with the place of human, and more particularly of individual and corporate subjectivity in new materialism. Undoubtedly, one of the main critical pushes of new materialism is a critique of individualism, the privileging of the human subject, of which the new materialists (quite rightly) find lingering vestiges in constructivist positions that dominated the cultural and semiotic turn of late-twentieth-century philosophy and theory. New materialists reject the illusion of the masterful agential subject who dominates over inert matter and, instead, want to speak of the "vitalism," even the "enchantment,"[55] of matter. However, not only is the human no longer the discrete center of agency but, as N. Katherine Hayles more radically puts it, "our saturation with networked and programmable media shunts us out of the realm of the human and into the realm of the posthuman: 'an informational pattern that happens to be instantiated in a biological substrate.'"[56]

While some authors in this volume are sympathetic to this Deleuzian view of the human as a dynamic "assemblage" without fixed ontology (Crockett and Reader), others (most notably Rieger, Lamothe, and Bray) express their concern that it may occlude the continued oppression of those subjects whose materiality is already most often rendered invisible. On the one hand, the eagerness to rush headlong into a post-human world signals a fatigue and frustration with a human species that proves itself again and again to be uniquely determined to out-do itself in greed, self-willed ignorance, laziness, and apathy. On the other hand, one can see this as a strategic weakness to posthumanism. Surely, humans are in as precarious a position as the other life forms of our planet, even if, unlike other species, that precarity is of our own doing. At a time in which real bodies are tortured, raped and mutilated, oppressed, starved, buried alive in collapsing clothing factories, cyber-bullied to the point of taking their own lives, driven from their homes in unprecedented numbers, economically repressed, deported, stopped and frisked, or even shot and killed for no other crime than the color of their skin, the academic rush to deny the identity, substantiality, and endurance of those human subjects surely can be challenged, as some of the following essays do.

LaMothe articulates how, in an attempt to shift away from anthropocentrism, the new materialists are in danger of "dissolving the materiality of a human bodily self into the flux and variability of matter in general." Rieger quite rightly raises the question of the identity of the "agent" in new materialism. In critical engagement with Crockett's and Robbin's version of materialism that wants to "posit earth as subject," Rieger suggests instead that we also ensure that we "add working people who are doing most of the work that sustains humanity and its progress as subjects as well." If the "human" is nothing more than a inchoate and random conflation of a myriad of material forces and powers, then whence comes the intentionality, let alone the sovereign power, to make a decision to act, and more than that, to resist? Such activism motivates the heart of the new materialist project, yet it may be pragmatically undercut by too much emphasis on the need to flee the subject. Minister's essay intervenes helpfully on this point with his nuanced analysis of the agency of conversions and the complexity of group orientations.

A second question is connected to the first and has already been mentioned: the status of the transcendent. Reading any of the new materialist writings, the emphasis on flux, dynamism, and the slipperiness of the material is readily apparent. One of the things that is clear in the rhetoric of new materialism is the need to emphasize the undecidability of the material, the amorphous shifting network of forces, the repulsion of static understandings of causality, and therefore the possibility of material agency. The elusiveness of matter goes hand in hand with the reasons for the rejection of transcendence. In both cases, it comes from an attempt to avoid authoritarian totalities. It is also seen as dangerously idealistic. Yet, as Rieger argues, transcendence need not be conceptualized as a "dangerous abstraction from real life" or solely in opposition to immanence. Further, as LaMothe points out, this rejection of transcendence also evacuates one of the primary motivators of our ethical action. She argues that it is important to ask "how our acts orient our senses" and what "relationship to the earth do they express... under the mantle of transcendent ideals," because such an appeal "enables us to know... [with every movement] we are participating in the creation of what is"; the appeal to the transcendent remains, for LaMothe, just that—an appeal rather than an acquisition. Its function is to add gravitas to our endeavors, but it can never act as a guarantor of them.

Another way to focus on this tension is to ask whether or not we can truly abjure subjective sovereignty if our framing remains immanent. In other words, do the central two claims of new materialism as a post-human and immanent framing not subtly work against one another? Does it make any difference if transcendence is posited not as guarantor or origin of system of thought and being, but more radically as an unknown hope for disruption, or excess of what we know, control, and manage?

This volume includes essays that advocate radical immanence as well as those which remain more agnostic about transcendence, challenging new materialism precisely when it stands in danger of becoming altogether too dogmatic about its insistence upon a radical immanence. All agree, however, that even if transcendence makes an appearance—either as a projected or hoped for horizon, or as a transgressive intrusion—it appears materially, immanently, objectively.

Thus far, many new materialists have simply ignored religion, despite the awareness that the religious lives of human subjects cannot be separated from the material worlds in which all beings move. This volume is an attempt to redress this lack. As Rieger points out, this approach will result in far more work for the scholar and require the ability to accept a "certain messiness." LaMothe also gestures to the need for the scholar of religion to be vulnerable—specifically, "to cultivate a vulnerability to be moved by those movements" the scholar witnesses. Above all, as the following essays remind us, new materialist approaches to the study of religion find their common ground in the insistence that religious experience is never purely interior and private, for it is writ in our actions, our orientations, and turnings, marked on our bodies and our expulsions, generative of energies and social movements that track one way or another. These are the clues that all new materialist scholars of religion read; and when they read these clues, they are also looking for something more, for ways to open up or join in other modes of becoming: the movements (of) matter. And this trajectory is something all scholars of religion, new materialist or not, should consider.

Notes

1. Sonia Hazard, "The Material Turn in the Study of Religion," *Religion and Society: Advances in Research* 4 (2013): 58–78. Accessed July 10, 2014. doi:10.3167/arrs.2013.040104.
2. For an explicit instance of such a call see David Chidester, "Material Terms of the Study of Religion," *JAAR* 68:2 (June 2000): 379.
3. Clayton Crockett and Jeffrey W. Robbins, *Religion, Politics and the Earth: The New Materialism* in *Radical Theologies Series* (New York: Palgrave Macmillan, 2012).
4. Diana Coole and Samantha Frost, eds., *New Materialisms: Ontology, Agency and Politics* (Durham, NC: Duke University Press, 2010); Rick Dolphijn and Iris van der Tuin, eds., *New Materialism: Interviews and Cartographies* (Ann Arbor, MI: Open Humanities Press, 2012).
5. Dolphijn and Tuin, *New Materialism*, 16.
6. See Stacy Alaimo and Susan Hekman, eds., *Material Feminisms* (Bloomington: Indiana University Press, 2008); Linda Alcoff, "Philosophy Matters: A Review of Recent Work in Feminist Philosophy," *Signs: Journal of Women in Culture and Society* 25:3 (2000): 841–882; Karen Barad, "Posthumanist Performativity: Toward an Understanding of How Matter Comes to Matter," *Signs: Journal of Women in Culture and Society* 28:3 (2003): 801–831; Rosa Braidotti, *Nomadic Subjects: Embodiment and Sexual Difference in Contemporary Feminist Theory* (New York: Columbian University Press, 1994); Elizabeth Grosz, *Volatile Bodies: Toward a Corporeal Feminism* (Sydney: Allen and Unwin, 1994); Sara Ahmed, "Open Forum Imaginary Prohibitions: Some Preliminary Remarks on the Founding Gestures of the 'New Materialism,'" *European Journal of Women's Studies* 15:1 (2008): 23–39.
7. See Leon Niemoczynski, "21st Century Speculative Philosophy: Reflections on the 'New Metaphysics' and Its Realism and Materialism," *Cosmos and History: The Journal of Natural and Social Philosophy* 9:2 (2013): 14.
8. Bruno Latour, *Pandora's Hope: Essays on the Reality of Science Studies* (Cambridge, MA: Harvard University Press, 1999); Bruno Latour, *We Have Never Been Modern*, trans. Catherine Porter (Cambridge, MA: Harvard University Press, 1993); Catherine Malabou, *The Future of Hegel: Plasticity, Temporality, Dialectic* (New York: Routledge, 2004); Catherine Malabou, *What Should We Do with Our Brain?* (New York: Fordham University Press, 2008); Quentin Meillassoux, *After Finitude: An Essay on the Necessity of Contingency*, trans. Ray Brassier (London: Continuum, 2008).

9. Niemoczynski, "21st Century Speculative Philosophy," 19.
10. Coole and Frost, *New Materialisms*, 5.
11. This classical view of nature and matter lent an erroneous "sense of mastery bequeathed to the thinking subject: the *cogito* (I think) that Descartes identifies ontologically other than matter... The Cartesian-Newtonian understanding of matter thereby yields a conceptual and a practical domination of nature as well as a specifically modern attitude or ethos of subjectivist potency" (Coole and Frost, *New Materialisms*, 8).
12. See Coole and Frost, *New Materialisms*, 9: "For materiality is always something more that 'mere' matter: an excess, force, vitality, relationality, or difference that renders matter active, self-creative, productive, unpredictable."
13. See Coole and Frost, *New Materialisms*, 11: "subatomic behavior consists in the constant emergence, attraction, repulsion, fluctuation, and shifting nodes of charge: which is to say that they demonstrate none of the comforting stability or solidity we take for granted."
14. Coole and Frost, *New Materialisms*, 15.
15. See for instance, Coole and Frost, *New Materialisms*, 7: "An important characteristic shared by all three components is their emphasis on materialization as a complex, pluralistic, relatively open process and their insistence that humans, including theorists themselves, be recognized as thoroughly immersed within materiality's productive contingencies."
16. In the case of Coole and Frost, or Dolphijn and van der Tuin, they are edited anthologies with multiple, often disagreeing voices represented; in the case of *Religion, Politics and the Earth*, while a single monograph, it is co-authored by Crockett and Robbins, with contributions from scholars outside of their domain (Michael W. Wilson and Kevin Mequet).
17. Coole and Frost, *New Materialisms*, 37.
18. Jane Bennett, "A Vitalist Stopover on the Way to New Materialism," in *New Materialisms: Ontology, Agency and Politics*, ed. Diana Coole and Samantha Frost (Durham, NC: Duke University Press, 2010), 47; emphasis added.
19. Jane Bennett, *Vibrant Matter: A Political Ecology of Things* (Durham, NC: Duke University Press, 2010).
20. Bennett, "A Vitalist Stopover," 47.
21. See Crockett and Robbins, *Religion, Politics and the Earth*, xx.

22. The concept of "assemblage" articulated by Deleuze and Guattari is helpful in this regard, especially in giving theorists a way to think beyond the subject/object divide. Rather than assuming there are discrete realities or objects that a discrete single knowledge apprehends, new materialists emphasize the co-constitutiveness of all things in vastly complex and dynamic "assemblages." See Gille Deleuze and Félix Guattari, *Anti-Oedipus: Capitalism and Schizophrenia*, trans. Robert Hurley, Mark Seem, and Helen R. Lane (Minneapolis: University of Minnesota Press, 1983).
23. For an excellent "materialist history" outline of the imbrications between geopolitics, the possibility of the development of culture, and the production and exploitation of cheap sources of energy, see Crockett and Robbins, *Religion, Politics and the Earth*, Chapter 6, "Energy." They summarize the impact of petroleum thus: "In short, the contemporary world is unthinkable apart from the revolution wrought by the cheap energy oil provides. Whether considering the rise of the Unites States as a global superpower, economic globalization, technological innovation, war, geopolitics, our contemporary world is one fueled by oil" (95).
24. Coole and Frost, *New Materialisms*, 5.
25. Coole and Frost, *New Materialisms*, 38.
26. Crockett and Robbins, *Religion, Politics and the Earth*, xv.
27. Crockett and Robbins, *Religion, Politics and the Earth*, xiv.
28. To give a small selection, see Crockett and Robbins, *Religion, Politics and the Earth*, xiii: "Witness the vapidity of most popular cultural modes: the saturation of consciousness by 'reality' shows... the bleeding of news into cynical infotainment, the dumbing-down or corporatization of education... While wealthy and not-so-wealthy Americans watch television and become zombies of consumption, the world's megacities are engaged in the 'mass production of slums'." In this context, they argue, democracy is a farce—pretty window dressing to distract from the special interests of the relatively few corporations it feeds.
29. Coole and Frost, *New Materialisms*, 25.
30. Coole and Frost, *New Materialisms*, 25.
31. See Coole and Frost, *New Materialisms*, 26: "to claim that something is constructed often has the unintended effect of recentering the human subject as the locus of agency despite the intention to undermine such claims. In other words, a constructivism that presumes matter's

passivity or plasticity in the face of power may echo an earlier ontology for which matter is inert stuff awaiting cultural imprint."

32. Coole and Frost, *New Materialisms*, 6.
33. Coole and Frost, *New Materialisms*, 6.
34. It is important to note, however, that this is one of those places where there is not agreement among new materialists. For instance, at least one prominent thinker attached to new materialism, Sara Ahmed, has explicitly argued that it does not entail a critique of constructivist theories of the subject. See Ahmed's contribution to Coole and Frost's volume, "Orientations Matter," in *New Materialisms: Ontology, Agency and Politics*, 234–257.
35. See in particular Dolphijn and van der Tuin, *New Materialism*, Chapter 6, "Pushing Dualism to an Extreme." Peta Hinton, in her review of the work, expresses it thus: the monism of New Materialism "shifts the status of matter from a primary ontological reality that precedes the properties of language, mind, subjectivity, and representation, to emphasize the relational and (therefore) durational dimensions of materiality in its entangled emergence with these ostensibly transcendent terms or substances" (Peta Hinton, *Hypatia Philosophy Review*, http://hypatiaphilosophy.org/HRO/reviews/content/195).
36. See Crockett and Robbins, *Religion, Politics and the Earth*, xxi: "We claim that energy is immanent Deleuzo-Hegelian spirit (or Spirit), and energy avoids the traditional dichotomy between spirit and matter because everything is energy-transformation."
37. Crockett and Robbins, *Religion, Politics and the Earth*, 79.
38. Crockett and Robbins, *Religion, Politics and the Earth*, 80.
39. Ludwig Feuerbach, *The Essence of Christianity*, trans. George Eliot (New York: Prometheus Books, 1989), 12.
40. Crockett and Robbins, *Religion, Politics and the Earth*, 18.
41. Crockett and Robbins, *Religion, Politics and the Earth*, 20.
42. When people quote Marx as saying that religion is the "opium of the people," they usually omit to put that statement into its full context. Immediately preceding that statement Marx writes, "Religion is the sigh of the oppressed creature, the heart of the heartless world, just as it is the spirit of a spiritless situation" (cited in Crockett and Robbins, *Religion, Politics and the Earth*, 21).
43. Crockett and Robbins, *Religion, Politics and the Earth*, 21.
44. Crockett and Robbins, *Religion, Politics and the Earth*, 22.
45. Feuerbach, *Essence of Christianity*, 13.
46. Crockett and Robbins, *Religion, Politics and the Earth*, 17.

47. In his chapter in this volume, Rieger picks up on Marx's critique of Feuerbach—that the point is not simply to analyze the conditions of our existence, but to change them—and pushes Marx to the next level in part by highlighting "the role of oppressed people as authoritative sources of religious experience." Furthermore, such religious experience might be motivation for liberation as well as exploitation.
48. Mark C. Taylor, ed., *Critical Terms for Religious Studies* (Chicago: Chicago University Press, 1998).
49. Robert A. Orsi, ed., *The Cambridge Companion to Religious Studies* (Cambridge: Cambridge University Press, 2012).
50. Robert H. Sharf, "Experience" in *Critical Terms for Religious* Studies, 107.
51. Robert A. Orsi, "The Problem of the Holy," in *The Cambridge Companion to Religious Studies,* ed. Robert A. Orsi (Cambridge: Cambridge University Press, 2012), 84, 90.
52. Orsi, "The Problem of the Holy," 92.
53. Crockett and Robbins, *Religion, Politics and the Earth*, 25.
54. Relational Christian realism is a progressive Christian movement coming out of the Church of England—though influenced by Reinhold Niebuhr, as much as William Temple—whose aims are the promotion of a more just and equitable society in balance with ecological concerns. The movement is primarily associated with the writings and activities of John Reader, Chris Baker, and John Atherton.
55. See Jane Bennett, *The Enchantment of Modern Life: Attachments, Crossings, and Ethics* (Princeton, NJ: Princeton University Press, 2001).
56. Coole and Frost, *New Materialisms*, 17.

Chapter 1

Becoming a Bodily Self
An Ecokinetic Approach to the Study of Religion

Kimerer L. LaMothe

In dance techniques across styles and genres, including western forms of ballet, jazz, and modern, dancers rely on a similar sequence of movements in order to launch their bodily selves off the ground and then safely, gracefully, land again. The sequence consists of several running steps followed by a leap. With each running step, a dancer builds intensity and speed into her bodily self, gathering up the energy she will release into the leap. On the final run, as her front foot touches down and presses into the earth, she swings her back leg forward to lift her whole bodily self off the ground. With this single motion of both pushing down and lifting up, a dancer propels herself through space, before she lands once again—moving up while touching down—on her outstretched, leading leg.

The two parts of this chapter each progress according to this run-run-run-leap logic. In Part One, with each running step, I analyze a central concern of new materialists shared by others in the field of religious studies as well: the role played by religious values in helping humans cultivate love for the earth—or not. The running culminates in a leap: I push off from new materialist positions to argue that a more promising role for religion in relation to the earth would appear if new materialists shifted their attention from *matter* to *movement*—as the enabling source and medium of life.

In Part Two, my second set of running steps collects resources for a concept of "bodily becoming" from three American modern dance pioneers, contemporary neuroscientists, and three "old" materialists. In a final leap, I lay out an *ecokinetic approach* to the study of religion in which "religion" appears as a primary activity within which humans learn to create and become patterns of bodily movement that help them engage their environments in life-enabling ways.

Part One: Love for the Earth

Run 1. New materialists want us to care about the earth. They want us to "believe" in the earth; to honor and love *Earth* as we honor and love ourselves, and thus, to bring a change to the way we treat it.[1]

In this concern with cultivating love for the earth, new materialists are not the first. Trailblazers in the American context include Henry David Thoreau and Aldo Leopold. As Leopold writes, while we may wax philosophical about environmental ethics, we will not act accordingly unless we have *love for the land*.[2] More recently, writers ranging from Wendell Berry to eco-feminist philosophers have picked up this theme, arguing that all the statistics in the world will not move us to act unless we *care* enough to do so.[3] As the title of Berry's Jefferson Lecture confirms, "It all turns on affection."[4] Joining this chorus, then, new materialists ask: *how can we encourage a belief in earth, a love for earth, a desire to act otherwise in relation to earth, and so save earth—and ourselves?*

While not new in their concern for the earth, new materialists enter the conversation intent on bringing new resources to it—resources offered by "old" materialists in the modern western tradition. Specifically, new materialists look to these older materialists for support in unmasking notions of transcendence—including ideas of God, Reason, the Self—as patterns of thinking that devalue the earth and its bodies. According to new materialists, the older materialists demonstrate how religion is—or at least has been—a preeminent form of earth-denying idealism. Religion is a form of wish fulfillment (Ludwig Feuerbach); it represents a fear of nature (Sigmund Freud); it serves as an opiate that masks the material conditions of suffering (Karl Marx), and mandates hostility to embodied, aesthetic life (Friedrich Nietzsche). Picking up these critiques, new materialists argue that the beliefs and practices of religion train human attention away from lived, bodily experience, away from the true sources of our suffering, and away from the healing potential of our own bodily becoming. As new materialists Clayton Crockett and Jeffrey Robbins relate, religion, in both philosophy and practice, "constructs an ideal world that we wish to inhabit and then mistakes that world for the real one."[5] This error, as Crockett and Robbins demonstrate, condones patterns of thinking and acting that are plunging us into ecological and economic crises.

Even as new materialists appropriate older materialist critiques of religion, they want to mobilize these critiques in order to demonstrate how religion can be useful for those interested in promoting ecological awareness, albeit in a negative way. By revealing religion as the practice of constructing ideal worlds, new materialists hope to harness the constructive potential it represents to catalyze social and political change. By unmasking appeals to transcendence as expressions of human desire, new materialists aim to free the love and attention we send up to God and divert it back down to earth.

Run 2. Among new materialists, however, this project of unmasking and defusing transcendent ideals never goes all the way. Most admit that the attempt to reduce all phenomena to material reality is neither possible nor desirable. Most notably, such a project loses the ability to affirm the agency and responsibility of human beings in relation to "matter." Without a human subject, a new materialist project is itself impossible: who will be there to care about the planet? Who will do the work involved in making a difference? Love for the earth requires someone who can do this work. New materialists, then, seek to achieve a "post-secular" stance. They aspire to a position that neither presumes nor reinforces a dualism between "matter" and "spirit" as ideal. They write about materiality in ways that leave room for affirming the movement of human subjects.

Here is where the new materialist drive to foster love for the earth gets interesting—and murky. More often than not, "religion," initially sidelined, slides back in. References to "the spiritual" and "the divine" surface in the form of the very subjective, sensory experiences that sustain human agency and will. At the same time, it remains unclear what roles religion can or should play in cultivating those experiences.

Take Rosi Braidotti, for example, the feminist philosopher credited with coining the term "new materialist." As she says in a recent interview, "'Neo-materialism' emerges as a method, a conceptual frame and a political stand, which refuses the linguistic paradigm, stressing instead the concrete yet complex materiality of bodies immersed in social relations of power." Here Braidotti targets the linguistic turn of postmodern scholarship evident in the 1990s and its attendant relativism as forms of earth-denying idealism. Braidotti aims instead to foster "robust alternatives" that deal with "issues of embodiment and accountability."[6]

In the name of fostering alternatives, Braidotti embraces a "concrete yet complex" notion of an all-encompassing matter that is internally diverse, dynamic, and always in the process of its own becoming. As she affirms, "The notion of the univocity of Being or single matter positions difference as a verb or process of becoming at the heart of that matter. There are only variations or modulations of space and time."[7] Thus, Braidotti rolls any sense of or appeal to a transcendent other into the single whole that unites us all. Braidotti insists that this new materialist move frees individuals to engage their lived bodily experience in reconceiving what they can and want to know, and what they hope may be. It is when Braidotti explains why an individual would want to engage in such a project that "religion" creeps back in. As she writes, new materialism "is a post secular position . . . [it] expresses sizeable doses of residual spirituality in its yearning for social justice and sustainability."[8]

Braidotti's intellectual moves are clear. (1) She affirms that new materialism does not traffic in transcendents (as do idealisms, including the linguistic paradigm); it affirms radical immanence.

(2) She qualifies that this immanence is not simple (as in, a reductive scientific or "old" materialism); it is internally diverse. (3) She affirms that, as post-secular, this "matter" moves beyond the spirit/matter binary to include "generous bonds" of community, a multiplicity of socially entangled bodies, and a place for the spiritual as a "residual" yearning.

It is worth noting, however, that this residual spirituality is not incidental to Braidotti's project. It is crucial: it funds the desire for social justice and sustainability that propels new materialism in the first place. As Braidotti affirms, the only reason to engage in the new materialist work of imagining robust alternatives to social constructivism (and other transcendent appeals) is *love*. As she writes, "Why should one pursue this project? For no reason at all. Reason has nothing to do with this. Let's just do it for the hell of it and for love of the world."[9] In Braidotti's account, religion lingers in the form of emotional patterns—shapes of sensation that swell within us, sweep through us, and spill from us. It appears within us as an imaginative act of reaching beyond the comfort zone of scientific reason, as the courage to feel present pain, as a spur to move otherwise. In this view, the love we need for the body of earth—and for our bodies of earth—is left over from a religion out of whose sphere we are nonetheless moving.

At this point questions arise. Of what exactly is this yearning a residue? How, where, and why does this yearning manifest in us? How do we know we can trust it, even while discarding other aspects of religion? And how do we keep alive the love and yearning that will sustain earth-friendly action, if not by engaging in the bodily practices of which that love and yearning are sensory affects?

Bill Connolly presents a second example. In an essay for *New Materialisms*, his stated aim is to elaborate upon a philosophy of immanence, that is, "a philosophy of becoming in which the universe is not dependent on a higher power."[10] He continues, with signature new materialist moves, by asserting that his philosophy "is reducible to neither mechanistic materialism, dualism, theo-teleology nor the absent God of minimal theology." Why not? As Connolly explains, it is because his notion of matter is complex: "there are uncertain exchanges between stabilized formations and mobile forces that subsist within and below them."[11] Nevertheless, like Braidotti, he also relies on a notion of "the spiritual" to leaven matter's mass. As he affirms, this complex matter harbors a "spiritual dimension of life" that not only arises as a desire for a philosophy of immanence, but serves as its enabling condition.[12]

In his effort to describe where this spiritual dimension resides within the dynamic complexity of matter, Connolly, like Braidotti, appeals to sensory experience. Specifically, he relies on an analysis of visual perception. Noting that visual "perception requires a prior *disciplining* of the sense," he acknowledges the implication—perception is vulnerable to persuasion.[13] Persuasion occurs through a process of disciplining, whereby the potential of our bodily selves to see is developed in relation to particular stimuli.

The fact that our senses can and must be disciplined gives Connolly hope. By writing, he intends to help people educate their senses to new ranges of experience—most notably, this spiritual dimension of material life. As he affirms, his philosophy aims to "open the anticipatory habits and sedimented dispositions of more constituencies during a time when media politics diverts attention from the most urgent dilemmas of the day." He too wants robust alternatives. He promotes a sensory education that will "deepen" our "visceral attachment to the complexity of existence"[14] and thus nurture our ability to love the earth—or, citing Deleuze—our ability to believe in this world.[15]

In advancing his account, Connolly again relies on sensory qualifications to distinguish the "spiritual dimension" of life from the mental constructs of "religion." As he clarifies, the "belief" he evokes does not involve adherence to a particular religious doctrine. Belief is more akin to a "vague existential disposition"—a "surplus" of meaning that dwells in the sensory space of our bodily selves. Belief, he explains, "touches... the tightening of the gut, coldness of the skin, contraction of the pupils, and hunching of the back that occur when a judgment or faith in which you are deeply invested is contested, ridiculed, ruled illegal, or punished more severely yet. It also touches on those feelings of abundance and joy that emerge whenever we sense the surplus of life over the structure of our identities."[16] Again following Deleuze, Connolly aims to attach this sensory surplus to "positive political movements that embrace minoritization of the world." He concludes that we need such ways of cultivating and mobilizing our energies because "a spirituality of some sort or other is always infused into experience, interpretation, and action."[17]

For Connolly, as with Braidotti, the spiritual returns as an originary impulse and enabling condition of new materialism itself: the spiritual is a viscerally felt commitment that impels us to think, feel, and act in love for the earth. As with Braidotti too, this series of moves raises questions concerning the relation of our bodily selves to the spiritual dimension of life. How are we supposed to discipline our senses to ensure that our "sensory surpluses" move in compassionate, sympathetic alignment with the environmental, social, and political crises of our time? How do we "cultivate and mobilize our energies"? Is there any role here for "religion"?

A final example appears in Clayton Crockett and Jeffrey Robbins's book *Religion, Politics, and Earth: New Materialism*. Crockett and Robbins also profess their desire for a "reenvisioned materialism"—one that counters any "idealism in its practical and philosophical forms."[18] They too make a second new-materialist move, refusing to reduce everything to natural laws. With allusion to Deleuze, Guattari, and, by extension, Spinoza, these authors claim that matter is planes and fields, gradients and intensities, pleats and folds. Matter is fertile, engendering, agential, and vibrant.[19] With these moves, Crockett and Robbins aim for "a deformation of theology and religion"; they aspire to "a new way of valuing life and the material world," including a perspective on religion that is "less dismissive and more radical."[20]

To launch this earth-friendly, less-dismissive deformation of religion, these authors revisit Feuerbach, Marx, and Freud, among others. "Religion," they assert, "is not about the truth . . . [It] is powered by the energy of desire"; it is "false consciousness."[21] Yet, Crockett and Robbins also insist that religious claims, because they answer to deep human needs, represent a force for change—a force that these authors want to harness in addressing the ecological, energy, and financial crises of our time.

Nevertheless, by their account, the social efficacy of religion rests on our intellectual understanding of how religious beliefs and practices perpetuate false consciousness. If we attend to religious beliefs, these authors aver, we can better understand "the proper structure of disavowal."[22] That is, we can recognize the ways in which we humans continually manufacture and project ideals of our own making in response to psychological and social needs. We can recognize the ways in which the actions that we take in our lives allow our beliefs to become real—to become knowledge—for us. In this way, "Religion is used to explain humanity to itself."[23] In later chapters, the authors suggest that this critique of religion can empower people to take charge of their own lives; revive their awareness of their inherent political power; resist the social and psychological pressures of digital, corporatized capitalist culture, and thus, devote themselves whole-heartedly to the discovery and implementation of earth-loving energy solutions.[24]

In Crockett and Robbins's effort to deform "religion," as in the work of Braidotti and Connolly, allusions to "spirit" erupt in places where the authors need to distinguish their notion of matter from the merely material. As they write, their materialism is "based on energy transformation. Energy... is not reductive matter but resonates with 'spirit' and 'life.'"[25] Similar questions arise. What is this "spirit," and how, where, and why does this "spirit" resonate? How is this resonance related to those deep needs to which religion supposedly responds? And once we reveal the structure of disavowal in religion, how are we to address these needs? Does religion still "work"? Are we supposed to convince ourselves to go through the motions of religious life, knowing that the point is to make real what we want to be true?

Run 3. In each of these three cases, new materialists recoup a notion of "spiritual" that they claim is not simply opposed to "material," but rather reverberates within it as a residual yearning, a sensory surplus, or a convenient fiction, respectively. Religion—initially dismissed as earth denying—lingers in perceptual habits, or patterns of sensation and response, that safeguard the willing, responsible core of a human subject. Its sensory artifacts subtend our desire for political action. At the same time, however, these authors' accounts of matter-rich-with-spirit leave little room either to affirm a responsible engagement with the *practice* of religion, or to frame the *study* of religion as much more than an exposing, disclosing critique.

Moreover, all three of these attempts to evince a post-secular, spiritually enhanced, earth-friendly politics pause in the same place: right when it comes to understanding how and why "religion" persists within our specific, sensory, bodily selves. While insisting on matter's complexity and dynamism across fields, surfaces, and planes, each of these theorists effectively dissolves a human "body" into a larger flux, without taking account of the physical particularity of our admittedly enmeshed, relational bodily selves.[26] While these new materialists all agree that the notion of matter with which they are operating is not your ordinary mechanical, Newtonian ideal, but complex, dynamic, and ever in the process of its own becoming, they are hard-pressed to identify a materiality specific to a human body. They are hamstrung in these efforts by their professed commitment to dissolve the individual, anthrocentric subject and his transcendent god.[27]

Take Crockett and Robbins as one example. After affirming that thinking is thoroughly material, in and of the body, they write, "Thinking returns to itself in and as earth, which is neither the reduction to crude materiality nor the embrace of mystical new age spirituality. This downward trajectory of thinking involves the spreading of sense back down into body and bodies, the body of earth, in its search for a new and vital event, one that will allow humans to survive and continue to think."[28] Here, thinking plummets from its abstract peaks, right down through "the body," and spreads out across the body of earth without stopping to take one lung-shaped breath. Similarly, elsewhere, after insisting that all thought is thoroughly material, Crockett and Robbins beat a hasty retreat from "the body" itself: "We must be clear, we do not mean this in terms of a crude reductionism, but rather as an embrace of the complexity of material organization and the further complexity and self-organization of brains, minds, and thoughts."[29] Back to minds and thoughts; a body, as a material thing, is revealed as what it "really" is: part and parcel of a larger dynamic field of becoming—which is, for Crockett and Robbins, best described as "a brain."

Leap. If new materialists want to find in religion—its beliefs and practices, values and experiences—an earth-friendly ally, they need to go further in attending to its bodily dimensions. They need a philosophical account of the specificity of human bodily selves and how these selves incorporate or resonate with a spiritual dimension of life. Specifically, they need an account of movement specific to human bodily selves that both relates and distinguishes those selves within a larger relational flux. They need an account of how and why *the movement of a human bodily self matters.* In brief, to make such a leap, new materialists need to move beyond *matter* (regardless of how dynamic and complex) as the primary locus of their analysis to focus on *movement* itself as the source and telos of life.[30]

A perceptual, conceptual shift from matter to movement is subtle, a flick of an eye. It does not involve looking elsewhere, but looking again. It requires that new materialists allow novel patterns of perception to come into focus by mobilizing different sensory patterns within themselves. In particular, it involves scholars activating a sensory awareness of a bodily self's movement-making potential—including their own.

When scholars understand that bodily selves *are* movement, that this movement occurs as a *rhythm* of creating and becoming *patterns of sensation and response,* and that humans *can* develop the capacity to participate consciously in this rhythm, then "religion" appears within human history as a place where humans learn *how* to develop this capacity. With this shift, new resources emerge for conceiving spiritual and material dimensions of bodily life as interrelated expressions of an ongoing *rhythm of bodily becoming.* This shift in perspective, then, would afford new materialists additional resources for participating critically in a process by which religious traditions may become allies in a global project of honoring the earth.

For help with making this shift, Part Two of this chapter turns to some unlikely and likely sources—making three more running steps and a final leap. The first two runs draw from the work of American modern dance pioneers and contemporary neuroscientists for help in fleshing out a concept of bodily becoming. A third run rereads the writings of three older materialists in light of this concept of bodily becoming. Attending to patterns of bodily movement evident in these materialists' critiques of religion unearths constructive, creative moments of their projects that have so far been overlooked, including guidance on how to create new values that can and will, in the words of Nietzsche's Zarathustra, "remain faithful to the earth."[31]

These three running steps culminate in a final leap: an ecokinetic approach to the study of religion in which "religion" names matrices of movement patterns within which people learn to create and become patterns of sensation and response—including patterns of believing, feeling, thinking, and acting—that will relate them to a given environment in life-enabling ways. Such an ecokinetic approach can provide scholars and practitioners, believers and skeptics with resources for evolving the movement patterns promoted by a given religious tradition in line with careful, actionable love for the earth.

Part Two: An Ecokinetic Approach

Run 1. The idea that a human being *is* a rhythm of bodily becoming first formed for me during a period of intense artistic and intellectual study dedicated to American modern dance pioneers, Ruth St. Denis, Isadora Duncan, and Martha Graham. As I have written extensively elsewhere, early in the twentieth century, these women were already revaluing religion into an earth-friendly ally through regular practices of rhythmic bodily movement.[32] Duncan and Graham, in particular, claimed to have discovered techniques of dance training capable of catalyzing a shift in experience for those who dance or watch—a shift in which a person could come to sense his bodily self as *movement*. When this experience-shift occurs, these women insisted, a dancer knows what he needs to know in order to hold any movements that he is making accountable to the ongoing health and wellbeing of the earth in him and around him.

Time and again, St. Denis, Duncan, and Graham describe how the act of dancing awakens a sensory awareness of one's bodily selves as inherently creative—as an endlessly generative source of movement patterns. As these women practiced the movement patterns they had discovered, they found new patterns of moving that opened out along the trajectories of attention and strength that their earlier movements had traced. Through this process, they came to know their bodily selves not just as movement, per se, but as ongoing *rhythms* of creating and becoming movement patterns or *kinetic images*. For these women, every human is such a *rhythm of bodily becoming*, and dancing is that action that exercises and makes this rhythm real, visible, and palpable both for those who practice and, vicariously, for those who bear witness.[33]

Further, in their practices of dance, Duncan and Graham sought not only to cultivate a visceral awareness of their inherent rhythmic creativity; they sought to develop an ability to participate consciously in this rhythm, so as to make movements that would grow their bodily selves in dance-enabling ways. Acutely aware of how every movement they made made them, these women set out to discern what their movements were creating in themselves and in the world. Key to this process were their own sensations of pain and pleasure. They strenuously rejected any technique that sought to impose movements upon a body, and instead looked for inspiration in a bodily self's natural, spontaneous movements—walking, jumping, running, leaping, breathing. They sought movements that would *honor and align with* the specific forms and capabilities of their own bodily selves. In doing so, these artists sought to create dances that would communicate participation in the love and attention to their bodily selves that such practice required.

In seeking to name their earth-honoring art, these dancers routinely called upon the language of religion. In Duncan's words, "The dance is not a diversion but a religion, an expression of life."[34] According to Graham, dance is "absolute knowledge,"[35] and "a revelation of experience."[36] St. Denis described her dancing self as a "rhythmic and impersonal instrument of spiritual revelation."[37] In choosing this vocabulary, these women were not claiming that their dances fit into known definitions of religion, nor were they referencing any specific tradition's beliefs. Rather, they were establishing that a human's conscious participation in the rhythms of bodily becoming—as practiced in dance—is an ongoing source of her highest ideals and values, her experiences and her aspirations.

Religion, these women averred, should be redefined in terms of what their dancing was and did.

Run 2. Recent discoveries in the sciences and social sciences are coming around to support what these modern dancers knew about the constitutive power of bodily movement. A primary catalyst for such studies occurred in 1996, with the discovery of mirror neurons. These neurons—located throughout the human brain in quantities larger than those of any other known primate's brain—fire when a person perceives patterns of movement. The pattern in which mirror neurons fire is the pattern that a perceiver would need to activate in himself in order to make the same bodily movement. Mirror neurons enable a human to move like an other, because of an other, in response to an other, and so create—in and as his own sensory self—a relationship with that other. As Ramachandran, a leader in the field, argues, mirror neurons "appear to be the evolutionary key to our attainment of full culture" for the way in which they allow us "to adopt each other's point of view and empathize with one another."[38]

These discoveries corroborate what the modern dancers taught and practiced: human beings *are* an ability to perceive movement patterns and recreate them in their sensory selves; and they have this ability to a degree unmatched in other primate or mammalian species. From conception to death, the process of sensing and responding to movement patterns is the means by which humans learn to do anything. Whether picking up a spoon, holding a pencil, grouping math variables, playing an instrument, or understanding Hegel, humans learn about themselves and the world by *recreating* in themselves the movement patterns that movements they have already made are enabling them to perceive. Whatever movements humans make educate and orient the senses, extending a person's powers of perception along the trajectories given by those movements. In the process, humans not only learn about themselves and the world, but they *become* themselves in relation to the world, even as the world becomes what it is in relation to them.

Studies in developmental psychology bring this point even closer to home when they draw connections between the movements humans make and their capacity to love. As researchers are finding, the bodily movements that infants and caretakers make in relation to one another—their ability to "move with" one another—predict the quality of their bond, or "attachment." Infants whose caregivers respond to their movements by recreating sympathetic shapes are more likely to develop the capacities of trust and empathy that long-term social relations require. A consensus is developing that our ability to attach with anyone or anything—that is, our ability to love—is a function of the bodily movements we are encouraged to make in our relationship with that other.[39]

As researchers across fields are thus observing, humans are first and foremost *rhythms of bodily becoming*. How humans move our bodily selves matters to the development of those qualities we prize as distinctively human—our empathy, compassion, cooperation, and self-consciousness. It matters to the feelings we feel, the thoughts we think, and the values we are able to hold. Moreover, as the modern dancers also discerned, this human capacity to sense and respond to movement patterns—our ability to "move with"—is not simply given; it is always already evolving as a result of the movements that are making us. As a result, through practice, humans can develop the capacity to discern how the movements we are making—whether freely or under pressure, carelessly or intentionally—as well as the movements that are being made by others in relation to us, are together making us into persons who think and feel and act in particular ways.

One implication, relevant to our present concern, is clear: if we are not engaging in movement practices that educate our sensory awareness to the movements that are making us, then we will not have the information we need in order to assess whether our ideals and aspirations, hopes and desires, are serving to create an earthly mesh of bodily relations in which we want to live. We will not have, within ourselves, the resources to move otherwise.

Run 3. Read through the lens of bodily becoming, the older materialists offer more guidance for the new materialist project than new materialists have gleaned to date. Feuerbach, Marx, and Nietzsche, as three examples, do not simply expose religious beliefs and practices as sundry forms of false consciousness. They affirm religion as an expression of a sensory creativity; religion's ideas and ethics represent attempts, both successful and otherwise, to perpetuate conscious participation in an ongoing rhythm of bodily becoming.

Take Feuerbach, whose *Essence of Christianity* functions as a "bible" of sorts for certain new materialists. As noted, new materialists embrace Feuerbach's critique that religion is a cultural moment in which humans project their own qualities onto a divine ideal. However, the new materialist focus on the material conditions of this projection diverts their attention from what Feuerbach identifies as the heart of the problem: the paralyzing effect that religious ideas can have on bodily movement. Reading Feuerbach from the perspective of bodily becoming, it is evident that Feuerbach evaluates religious beliefs and practices based on whether they promote conscious, earth-tuned participation in the rhythms of bodily becoming.

As Feuerbach sees it, the problem with religion is not that humans project ideals of themselves—humans cannot *not* do so. For Feuerbach, every human projects an aim, a purpose—an ideal image of herself that transcends her current situation—and, as such, has a "religion." This projection is not simply an ideal that a person *thinks* about her self. Any projection is an "impulse of life"—a *movement*—an action of moving out and beyond one's individual self. As Feuerbach writes: "Every man... must place before himself a God, i.e., an aim, a purpose. The aim is the conscious, voluntary, essential impulse of life... the unity of the material and spiritual in the individual man... He who has an aim... has... a religion."[40] To be human is to have or rather *to be* this "impulse"—this movement toward—which is as much material as spiritual, as much felt and wanted as it is conceived and thought. In making this point, Feuerbach confirms that humans *are* a dynamic process of creating and becoming images of their bodily selves. The impulses of life that manifest through human thoughts and desires—the bodily movements that humans make—contribute to making them who they are. Because humans are an ongoing, rhythmic movement of their own bodily becoming, Feuerbach affirms, humans cannot *not* have religion. Even his own philosophy, he avers, is "a religion."[41]

Nor is the problem that religious projections themselves—Christian ones in particular—represent false images of human being. On the contrary, what is represented in Christian teachings—according to Feuerbach—are *facts* of human being. Humans *are* thinking, loving, and willing creatures; and their thinking, loving, and willing *are* inherently creative, infinite, and thus divine. As Feuerbach affirms, when a human looks at God or into God, what he sees *is* his own potential—his humanity. "God is the mirror of man."[42]

Problems arise (with religion in general and Christianity in particular) when humans get stuck looking in the mirror. Humans make the movements of consciousness—projecting images of who they are—and then fixate on them as if those images existed outside of their own action as something "real." Dazzled by their own creations, they fail to see those images as *kinetic*—that is, as expressions of an "impulse of life"—as patterns of movement potential. As a result, they misunderstand what these images of themselves represent: neither "truth" nor "reality" per se, but guidelines for *moving* in distinctively *human* ways. In this reading of Feuerbach, humans err when they fail to perceive religious ideas as (what I call) kinetic images—patterns of sensing and responding, thinking, feeling, and acting, that humans can and should recreate in themselves to guide them in making movements that will realize their capacity to love themselves, others, and the earth.

When persons fail to understand religious beliefs and rituals as movement patterns, Feuerbach laments, they become "religious fanatics" who privilege their own aims above the wellbeing of other humans. Alternately, they become theologians who dissect and debate images of God, arguing over whether these ideas are right or wrong, as if ideas existed apart from the "essential impulse" of a whole human person in creating and becoming them. In either case, fanatics or scholars act in ways that prevent them from learning from religious ideas and practices how to move their bodily selves in ways that will realize the divinity of their own thinking, willing, and loving. For Feuerbach, the reason to study or practice religion is to *become more fully human*—that is, be able to participate consciously and responsibly in the act of creation, the impulse of life, occurring in us and through us in every moment, with every movement we make.

To support his case, Feuerbach references the Last Supper. As he points out, Jesus instructs his followers to eat and drink in memory of him. In other words, Jesus calls his followers to make the bodily movements that he is making—some of the most basic life-sustaining movements—in order to know who he is; in order to know the divinity of their own bodily selves. It is not that these actions make belief in God real. Rather, these actions of eating and drinking cultivate in a person a sensory awareness of her own bodily movements (in eating and drinking) as the medium in which to know the "religious import" of life. As Feuerbach explains, "Hunger and thirst destroy not only the physical but also the mental and moral powers of man; they rob him of his humanity—of understanding, of consciousness. Oh! If thou shouldst ever experience such want, how wouldst thou bless and praise the natural qualities of bread and wine, which restore to thee thy humanity, thy intellect! It needs only that the ordinary course of things be interrupted in order to vindicate to common things an uncommon significance, *to life, as such, a religious import*."[43] To recreate in one's own life the bodily movements that Jesus makes, to do so *for* him, while remembering him, is to cultivate a sensory awareness of God as present in "the natural qualities" of "common things." In this reading of Feuerbach, then, Christian ideas and practices work as they should when they engage a human capacity to perceive and recreate patterns of bodily movement so that humans can learn to exercise this power in ways that will connect them with others as Jesus did: in love. In a word, as the modern dancers would affirm, "divinity" (our own included) reverberates in a human capacity to *dance*.

Further, Feuerbach continues, once we make bodily movements that cultivate in us a sensory awareness of life's "religious import," we realize that divinity does not end with humans either. As Feuerbach urges, when we remember the significance of wine and bread, we realize our debt to "holy Nature." "Forget not that wine is the blood of plants, and flour the flesh of plants, which are sacrificed for thy well-being!"[44] For Feuerbach, the religious import of life is not a symbolic gloss applied to matter. Nor does the divine inhere in things. Rather, divinity flows in the movements that holy Nature is constantly making in relation to itself as it seeks its own ongoing generation—including those movements that occur in and through us.

In this reading, then, the task that Feuerbach sets for his readers is not to reject ideals that transcend our finitude (we cannot help having such ideals); nor is it to deny their reality (reality is not the point). The task is neither to dismiss religion as false consciousness, nor to study it in order to better understand the structure of disavowal. Rather, the task is to greet the beliefs and practices of religion as *kinetic images*—that is, as impulses of life, patterns of movement potential, that not only trigger a person's ability to create and become patterns of sensation and response, but guide one to do so in human-enabling ways. When we do, we will be more able to feel in and through our individual bodily selves the *love* that holy Nature is constantly practicing in relation to itself through us.

This approach to reading Feuerbach illuminates anew the responses that other older materialists make to him. Both Marx and Nietzsche, for example, may be read as articulating the challenges involved in creating and becoming *kinetic images* (e.g., religious beliefs and practices) that cultivate our human capacity to love one another and the earth. For both Marx and Nietzsche, external and internal factors distort the creative process by training human awareness away from the sensations of bodily movement. Yet humans need this sensory awareness in order to discern whether or not their religious ideas and practices are bringing forth an earthly world in which they can and want to live.

As Marx contends, a person's rhythms of bodily becoming occur in matrices of social relations that (depending on how those relations are arranged) can create obstacles to remembering and becoming divine. Capitalism, as a case in point, does mount such obstacles. How so? As Marx explains, the material conditions of capitalist production train a worker's sensory attention away from the concrete product he is making and toward the abstract wage he expects to receive.[45] His labor is no longer his. The product is not his; it confronts him as an independent or "alien" object. He cannot see in the product the bodily movements he made in making it. As Marx writes, "The alienation of the worker in his product means not only that his labour becomes an object, an external existence, but that it exists outside him, independently, as something alien to him, and that it becomes a power of its own confronting him."[46] This "labor," moreover, is not only bodily movement. It is the capacity to find and discover movement patterns that will align one's bodily potential with the requirements of the production process. As far as a worker participates in this economy, he is training his capacity for movement making to operate in the abstract realm of capital, beholden to someone else's bottom line.

What is created through the labor process, then, is not only a product and a profit; nor is it the body of the worker per se. What is created are patterns of movement that constitute the relationship of a person to her self, to others, and to the natural world. What is created is a person who senses and responds to her own movement-making potential as if it were an object, a machine, or a tool. For Marx, such a person is willing and able to believe in a power that transcends her individual self because such a belief repeats the movement pattern that she is creating and becoming in relation to her own body.

Here, Marx's comparison of religion to "opium" is apt.[47] Religious ideas and values operate at the level of *sensation*. They define patterns of attention that divert a person's sensory awareness away from his bodily self. They prevent a person from feeling what he is feeling. As a result, a person is unable to perceive the sensations of pain and pleasure that would otherwise guide him to move in life-enabling ways. Religion, as an opiate, works by preventing a person from participating consciously in the rhythms of bodily becoming. It denies workers the information they need to hold their own creative movement making accountable to the health and wellbeing of their bodily selves.

Thus, for Marx as with Feuerbach, the problem with religion does not lie in its appeal to the transcendent per se. Humans cannot *not* labor in the creation of products, values, and ideas that transcend their particular time and space. Problems occur when religious ideas and practices induce a kind of trance and erase sensation. By implication, religious consciousness (or religious experience) is not "false" relative to some "true" experience of oppressive material conditions. Religious consciousness is false insofar as the bodily movements prescribed by belief in god do not sustain the physical life of the bodily self who labors and prays. Religious consciousness is false relative to the earth—relative to the enabling bodily conditions of its ongoing production—like capitalism itself.

When Marx describes the critique of religion as "the premise of all criticism" on the path to social justice, then, he does so because this critique puts people in touch with their own movement making.[48] It galvanizes people to participate *consciously* in the rhythms of bodily becoming, ensuring that any labor they invest, any ideas or images they create, honor their ongoing ability to do so. In this way, the critique of religion is not a once-and-for-all act. It is a means for sustaining a "continuous intercourse" with nature that produces both objects and transcendent ideals. As Marx writes: "Man *lives* on nature . . . nature is his *body*, with which he must remain in continuous intercourse if he is not to die. That man's physical and spiritual life is linked to nature means simply that nature is linked to itself."[49]

Nietzsche's response and reworking of the materialist critique further illuminates a possible role for religion in promoting conscious, earth-friendly participation in the rhythms of bodily becoming. Nietzsche agrees with Feuerbach and Marx: humans do not *love* themselves enough—they do not love the *earth* enough.[50] He also agrees that humans are inherently creative, where their movement of creation is as physical as it is spiritual—an impulse of life. As noted earlier, in the words of Zarathustra, Nietzsche calls for humans to create highest values that "remain faithful to the earth." However, Nietzsche further insists that such value creation requires more than either a Feuerbachian mental shift or Marxist revolution. As he writes, humans need positive bodily practices—they need art—to seduce them away from the heavens and back to bodily life. They need to *cultivate* (and not just know about or assume) a sensory awareness of themselves as movement.

One of the many ironies of Nietzsche's legacy is that he is most often read as a mere critic—even a nihilist and relativist—even though the images of dance he uses in every major work outline a radical, constructive project.[51] As Nietzsche predicted, those readers whose senses have been colonized by verbal practices will tend to pass over these dance images, assuming that they are merely metaphors. Readers do so because their indoctrination in the production processes of reading and writing (as with Marx's capitalism) has trained their sensory awareness away from their own bodily movement making—away from the sources of their own creativity. For Nietzsche, belief that the power of reading and writing grants access to truth is one more manifestation of an ideal, earth-denying value. As he writes, "I am afraid we are not rid of God because we still have faith in grammar."[52]

Of course, Nietzsche does not advocate abolishing reading and writing all together. Rather, he advocates engaging in complementary bodily practices—a kind of "physical-spiritual discipline" or "morality"—that he describes from his earliest works to his final essays as "dance."[53] For Nietzsche dance is an *effective* symbol that points outside of the text to the kind of sensory work that a human bodily self must do in order to have the strength and discernment needed to create values that will remain faithful to the earth.[54] For Nietzsche, a person who "dances" is one who cultivates a sensory awareness of the movements that are making her. As Nietzsche demonstrates in *Ecce Homo*, such a "discipline" involves paying attention to how, when, and what we eat; when and what we read; how much we sit and whether we walk. It involves paying attention to which movements deplete us and which ones energize us; which depress our thoughts and feelings, and which enhance our sense of strength. It involves walking and being outdoors in the natural world.[55] It involves learning to make the bodily movements of Zarathustra, so as to learn the secret of how he "walks like a dancer."[56]

Further, this discipline also involves seeking out *art*—that is, expressions of human creativity that call our sensory attention into and through our bodily selves. For Nietzsche, the paradigm for an effective art is dance. As Nietzsche affirms, "All art works tonically, increases strength, inflames desire (i.e., the feeling of strength), excites all the more subtle recollections of intoxication—there is a special memory that penetrates such states: a distant and transitory world of sensations here comes back."[57] As these sensations come back, a person has what he needs to discern how to move in relation to all that is, so as to cultivate relationships that enable his "great health."[58]

While Nietzsche is infamous for the line he gives to a fictional madman—God is dead—a closer reading of his work reveals the reverse is true as well.[59] God is not dead; or rather, humans cannot *not* create gods who will too die someday. Humans cannot not create and become patterns of sensation and response in relation to what they value most. Yet, once humans cultivate a sensory awareness of themselves as movement, Nietzsche affirms, they will be able conceive of ideas and values that support and sustain this ongoing, conscious participation in the rhythms of bodily becoming. They will have within their own sensory awareness the moral compass they need to ensure that their highest values "remain faithful to the earth." As he asserts in the words of Zarathustra, "I would only believe in a god who could dance."[60]

Leap. This gathering of support from the modern dancers, scientists, and older materialists finds its release in a leap: an ecokinetic approach to religion. An ecokinetic approach perceives religion itself—its beliefs, practices, stories, and traditions—as patterns of movement made and remembered over time by those who have found these patterns helpful in creating relationships with the sources of their ongoing survival.

As seen above, this leap presses forth with the support of Feuerbach, Marx, and Nietzsche, who frame religion as a primary activity in which humans awaken and exercise a constitutive ability—substantiated by both modern dancers and neuroscientists alike: the human ability *to create and become patterns of movement—relational patterns of sensing and responding.* With this footing, religion appears as a nexus of kinetic images that choreograph the movements of projecting, laboring, imaging, and creating that a human essentially *is.* The kinetic images of religion are vital to humanity, moreover, not as claims to truth, but as moments along an educational path that make certain ways of distinctively human thinking, feeling, and acting possible. An ecokinetic approach thus confirms that activities that occur in religion are vital to human wellbeing, because it is in such activities—in the work of honoring, moving with, and manifesting imagined worlds—that humans learn to tap and exercise their inherent sensory creativity.

At the same time, as the modern dancers and older materialists avail, humans are all too ready to fix(ate on) the kinetic images of their own creation. If and when they do, humans arrest the "spiritual" in such a way that the ideals and values they create and become fail to honor the ongoing rhythm of bodily becoming from which such ideals and values emerged in the first place. In such cases, humans create and become patterns of sensation and response that authorize virtual and verbal worlds rather than nourishing the ongoing health of their relational, bodily lives. The implication, as dancers and materialists suggest, is that humans need to engage on an ongoing basis with practices of bodily movement that not only cultivate a sensory awareness of the movements that are making them, but also keep that sensory awareness open to novel impulses of life—impulses to move. In this way, such practices ensure that our highest values and aspirations remain faithful to the ongoing generativity of the earth—our continuous intercourse with holy Nature—as it is occurring in us, around us, and through us. Such practices do so, as the modern dancers knew, by enhancing our ability to perceive and respond to patterns of pain and suffering, oppression and despair, as instructions to move differently.

How can an ecokinetic approach help new materialists lift up and advance their accounts of religion so as to illuminate a role for religious ideas, values, and traditions in the project of stirring up love for the earth?[61]

First, as *kinetic*, an ecokinetic approach focuses a scholar's attention on the patterns of movement (eye, ear, and nose, hand, hamstring, and heart, intelligence, emotion, and spirit) that a given group of people remembers and repeats in the name of religion. These patterns may be handed down through generations or improvised on the spot; they may be public or private, rehearsed or spontaneous. An ecokinetic approach guides a scholar to discern how, why, when, and where people are moving because of, and in relationship to, what they claim has value or importance. It points to the ways in which people in a given context are exercising their movement-making potential and training their sensory awareness.

Second, an ecokinetic approach is *empathic* and, in this way, participatory. Once a scholar learns to perceive religious phenomena as movement patterns, an ecokinetic approach guides her to cultivate a vulnerability to being moved by those movements. Engaging her own kinetic imagination, she consciously recreates the patterns of movement in her own sensory awareness. As she recreates these movement patterns within her sensory self, she can feel and notice what further movements of thinking, feeling, and acting become possible. In this way, she is able to discern how the action of making these movements may train a person's senses to certain regions and areas of both pain and pleasure. While it is not necessary for a scholar actually to make another's movements in order to empathize with them, the versatility and range of her kinetic imagination will depend on the degree to which she has awakened in herself—through practices of bodily movement—a sensory awareness of participating in the rhythms of bodily becoming.

Third, an ecokinetic approach guides scholars to *evaluate* religious phenomena relative to the "eco" or ecological matrices of which humans are members. This evaluation is not an objective critique. The scholar as well as the phenomena he studies are equally enmeshed in the webs of relational nodes that allows them to exist—singly and together. As such, whatever the scholar does in the effort to study pulls on the web of relations within which he and the phenomena dwell. Yet, at the same time, any sensory act of attending or discerning is an act that inherently assigns value to that which appears. As a result, in an ecokinetic approach, a scholar cannot engage in the study of religious phenomena without asking, at every turn of his project *What am I creating as I engage with these phenomena?* He cannot not ask *How is my participation in the life of these phenomena affecting their ongoing trajectory of development?* A scholar *cannot* not ask *How can I engage in this critique such that I encourage whatever potentials for earth-friendly belief and action exist within the ongoing existence of "religion"?*

Again, this evaluative moment of an ecokinetic project does not involve bringing fixed criteria to bear on volatile phenomena. Rather, the evaluative moment is keyed to movement—to the kinds of movement potentials that are being open and closed in the moment. The criteria for evaluating the effects of religious belief and practice in this approach are constantly being generated within the act of studying as religion itself evolves. Do the religious beliefs and practices in question guide people to move in ways that educate their senses to the ongoing process of their own bodily becoming? Do these religious beliefs and phenomena fund a conscious participation in the work of creating and becoming life-enabling movement patterns? Do these religious beliefs and practices, in other words, serve to support the earthly conditions of their own ongoing relevance to human lives?

Insofar as religious beliefs and actions educate our attention away from the sensory awareness that humans need to ensure that our values remain faithful to the earth, then religion will inevitably (as Marx warned of capitalism) consume the sources of its own generative activity. People will get stuck looking in the mirror of their own projections (as Feuerbach warned); their values and ideals will express hostility to the earth, while endorsing and funding the life of something *other* than human wellbeing (as Nietzsche warned).

With an ecokinetic approach, however, we can appreciate the ways in which religious ideas and values may also fund our ongoing bodily becoming. We realize that "residual" yearnings for social justice (identified by Braidotti) signal trajectories along which religious traditions can and must evolve. We realize that we can and must adopt bodily practices that cultivate and direct our "sensory surpluses" (as Connolly suggests), and we do so when we engage in actions that pull our sensory attention through our bodily selves and exercise our capacity to participate consciously in the rhythm of bodily becoming. In such practices, we cultivate love for the earth—not as an object for which we care but as the very movement that is making us.

Conclusion

As Crockett and Robbins say, "If we want our civilization to live on a little longer we will have to recognize our coexistence with and in Earth which is not simply the ground of our being, but the dynamic intensity of our life and thought."[62] If we want to love the earth, we must love the earth in our bodily selves. If we want to move with the earth, we must quicken a sensory awareness of our relationship to it and of our participation in it. We must feel the pain of the earth as our own and find in that pain impulses to move otherwise. The value of doing so under the mantle of transcendent ideals is evident. It is not that we need to consult "god" about how we should move. Rather, by appealing to a transcendent, we allow ourselves to know viscerally, that, with every move we make, we are co-creators of the reality in which we live.

When we cultivate this visceral knowledge, our actions will spontaneously align more and more with our ever-evolving understanding of what must happen to sustain conscious participation in this rhythm of bodily becoming, and the ecological matrices it expresses. We shall know why we must *dance*.

NOTES

1. William E. Connolly, "Materialities of Experience," in *New Materialisms: Ontology, Agency, Politics*, eds. Diana Coole and Samantha Frost (Durham, NC: Duke University Press, 2010), 196. "New materialists" include a range of thinkers and writers across fields who reject appeals to transcendence in favor of a robust embrace of material, embodied life, here and now. For two recent anthologies, see: Diana Coole and Samantha Frost, eds., *New Materialisms: Ontology, Agency and Politics* (Durham, NC: Duke University Press, 2010); Rick Dolphijin and Iris van der Tuin, eds., *New Materialism: Interviews and Cartographies* (Ann Arbor, MI: Open Humanities Press, 2012).
2. Aldo Leopold, *Sand County Almanac* (New York: Ballantine Books, 1949), xix.
3. For a perspective on care, see Karen J. Warren, *Ecofeminist Philosophy: A Western Perspective on What It Is and Why It Matters* (Lanham, MD: Rowman & Littlefield, 2000).
4. Wendell Berry, *It All Turns on Affection: The Jefferson Lecture and Other Essays* (Berkeley, CA: Counterpoint, 2012).
5. Clayton Crockett and Jeffrey W. Robbins, *Religion, Politics, and the Earth: The New Materialism* in *Radical Theologies Series* (New York: Palgrave Macmillan, 2012), xvi.

6. "Interview with Rosi Braidotti," in *New Materialism: Interviews and Cartographies*, ed. Rick Dolphijn and Iris van der Tuin (Ann Arbor, MI: Open Humanities Press, Michigan Publishing, University of Michigan Library, 2012). http://quod.lib.umich.edu/o/ohp/11515701.0001.001/1:4.1/--new-materialism-interviews-cartographies?rgn=div2;view=fulltext.
7. "Interview with Rosi Braidotti."
8. "Interview with Rosi Braidotti."
9. Rosi Braidotti, "The Politics of 'Life Itself' and New Ways of Dying," in *New Materialisms: Ontology, Agency, Politics*, ed. Diana Coole and Samantha Frost (Durham, NC: Duke University Press, 2010), 217.
10. Connolly, "Materialities of Experience," 178.
11. Connolly, "Materialities of Experience," 178–179.
12. Connolly, "Materialities of Experience,"196.
13. Connolly, "Materialities of Experience," 187.
14. Connolly, "Materialities of Experience," 191.
15. Connolly, "Materialities of Experience," 196. See also Gilles Deleuze and Felix Guattari, *One Thousand Plateaus: Capitalism and Schizophrenia*, trans. Brian Massumi (Minneapolis: University of Minnesota Press, 1987).
16. Connolly, "Materialities of Experience," 196.
17. Connolly, "Materialities of Experience," 196–197.
18. Crockett and Robbins, *Religion, Politics, and the Earth*, xvi.
19. As Crockett and Robbins acknowledge, their concepts of matter are deeply indebted to Spinoza, Deleuze, and Guatarri. See Benedict de Spinoza, *A Spinoza Reader: The Ethics and Other Works*, ed. and trans. Edwin Curley (Princeton, NJ: Princeton University Press, 1994); Deleuze and Guattari, *One Thousand Plateaus*. For further discussion, see Jane Bennett, *Vibrant Matter: A Political Ecology of Things* (Durham, NC: Duke University Press, 2010).
20. Crockett and Robbins, *Religion, Politics, and the Earth*, xix.
21. Crockett and Robbins, *Religion, Politics, and the Earth*, 22.
22. Crockett and Robbins, *Religion, Politics, and the Earth*, 29.
23. Crockett and Robbins, *Religion, Politics, and the Earth*, 18.
24. Crockett and Robbins, *Religion, Politics, and the Earth*, Chapters 3–7.
25. Crockett and Robbins, *Religion, Politics, and the Earth*, xvi.
26. Braidotti, for example, describes "the body" as "a transformer of flows and energies, affects, desires and imaginings... It is the complex interplay of the highly constructed social and symbolic forces.

The body is a surface of intensities and an affective field in interaction with others" ("Interview with Rosi Braidotti").

27. "Anthrocentric" is a term used by Valerie Plumwood in *Feminism and the Mastery of Nature* (New York: Routledge, 1993) to represent both "anthropocentric" and "androcentric" perspectives. For a parallel critique of work by Deleuze and Guatarri, Jane Bennett, and David Abram, see Kimerer L. LaMothe, "Can They Dance? Towards a Philosophy of Bodily Becoming," *Journal of Dance and Somatics* 4:1 (2012): 93–107.
28. Crockett and Robbins, *Religion, Politics, and the Earth*, 140–141.
29. Crockett and Robbins, *Religion, Politics, and the Earth*, 132.
30. For a more extensive argument, see Kimerer L. LaMothe, *Why We Dance: A Philosophy of Bodily Becoming* (New York: Columbia University Press, 2015).
31. Friedrich Nietzsche, "Thus Spoke Zarathustra," in *The Portable Nietzsche*, ed. Walter Kaufmann (New York: Penguin, 1954), 125.
32. See Kimerer L. LaMothe, *Nietzsche's Dancers: Isadora Duncan, Martha Graham, and the Revaluation of Christian Values* (New York: Palgrave MacMillan, 2006).
33. LaMothe, *Nietzsche's Dancers*, Part II (for Duncan) and Part III (for Graham).
34. Isadora Duncan, *Art of the Dance* (New York: Liveright, 1928), 142.
35. Martha Graham, "A Modern Dancer's Primer for Action," in *Dance: A Basic Educational Technique*, ed. Frederick Rogers (New York: Macmillan, 1941), 180.
36. Martha Graham, "Preface," in *Martha Graham: Sixteen Dances*, ed. Barbara Morgan (Dobbs Ferry, NY: Morgan and Morgan, 1941).
37. Ruth St. Denis, *An Unfinished Life* (New York: Harper & Brothers, 1939), 52.
38. V. K. Ramachandran, *The Tell-Tale Brain: A Neuroscientist's Quest for What Makes Us Human* (New York and London: W.W. Norton, 2011), xv–xvi. See chapter 4 for discussion.
39. This idea is developed by Louis Cozolino in *The Neuroscience of Human Relationships: Attachment and the Developing Social Brain* (New York: W.W. Norton, 2007) and also by Daniel Siegel in *The Developing Mind: How Relationships and the Brain Interact to Shape Who We Are* (New York and London: Guilford Press, 1999).
40. Ludwig Feuerbach, *The Essence of Christianity*, trans. George Eliot (New York: Harper Torchbooks, 1957), 64.
41. Feuerbach, *The Essence of Christianity*, xliii–xliv.

42. Feuerbach, *The Essence of Christianity*, 63.
43. Feuerbach, *The Essence of Christianity*, 277–278.
44. Feuerbach, *The Essence of Christianity*, 277.
45. Karl Marx, "Economic and Philosophic Manuscripts of 1844," in *The Marx-Engels Reader*, ed. Robert Tucker (New York: W.W. Norton, 1978), 73.
46. Marx, "Economic and Philosophic Manuscripts," 72.
47. Karl Marx, "Contribution to the Critique of Hegel's *Philosophy of Right*: Introduction," in *The Marx-Engel Reader*, ed. Robert Tucker (New York: W.W. Norton, 1978), 54.
48. Marx, "Contribution to the Critique," 53.
49. Marx, "Economic and Philosophic Manuscripts," 75.
50. Nietzsche, "Thus Spoke Zarathustra," 173.
51. For a survey of these images and an interpretation of their significance in relation to Nietzsche's project of revaluing values, see LaMothe, *Nietzsche's Dancers*, Part I.
52. Nietzsche, *Portable Nietzsche*, 483.
53. Friedrich Nietzsche, *Will to Power*, trans. Walter Kaufmann and R. J. Hollingdale (New York: Vintage, 1968), #981, 513.
54. LaMothe, *Nietzsche's Dancers*, Chapter 3.
55. Friedrich Nietzsche, *On the Genealogy of Morals and Ecce Homo*, ed. and trans. Walter Kaufmann (New York: Vintage, 1989).
56. Nietzsche, "Thus Spoke Zarathustra," 122.
57. Nietzsche, *Will to Power*, #809, 427.
58. Friedrich Nietzsche, *The Gay Science with a Prelude in Rhymes and an Appendix of Songs*, trans. Walter Kaufmann (New York: Vintage, 1974), #382, 346.
59. Nietzsche, *The Gay Science*, 181.
60. Nietzsche, "Thus Spoke Zarathustra," 153.
61. As the following three criteria suggest, an ecokinetic approach engages and extends the phenomenology of religion in ways that attend specifically to the appearance of movement patterns in relation to what appears as "religion." For the roots of this approach in the work of Gerardus van der Leeuw, see Kimerer L. LaMothe, *Between Dancing and Writing: The Practice of Religious Studies* (New York: Fordham University Press, 2014), Chapters 5 and 6.
62. Crockett and Robbins, *Religion, Politics, and the Earth*, 154.

Chapter 2

Organizing Bodies

Theorizing Religious Conversion with New Materialisms

Kevin Minister

For many religious traditions, "conversion" names a foundational religious experience, definitive of what it means to be part of the religious tradition. However, discussions of religious conversion have fallen on hard times as a result of the association of conversion with colonial expansionism, social exclusion, and religious conflict. Even when the pursuit of religious conversion explicitly rejects violent means of coercion, conversion has often carried with it the implicit submission to dominant cultural, political, and economic systems.[1] The experience of conversion in this manner facilitates the expansion of cultural structures including language, gender, race, ethnicity, political orders, and economic systems.[2] Yet, conversion remains significant to religious identity and the ability to respond to the colonial history of religious conversion requires conversions that resist material, cultural, and political domination. Religious studies and theology need new ways of conceptualizing conversion that are responsive to the material, cultural, and political dimensions of conversion and attend to forms of religious conversion that resist domination.

Religious relativism provides a popular solution to the potentially violent contentions of conversion. Religious relativism individualizes and morally equates all choices about religious identity, thereby conceptually limiting conversion to private, personal decisions without public significance. While religious relativism may appear to fit well with dominant strains of American Protestant Christianity that focus on individual belief as the core of religious identity, religious relativism does not account for the numerous traditions of religious conversion that are more concerned with the formation of public or social practices in religious identity. As a result, religious relativism threatens to downplay the history and continuing potential of religious conversion to promote or sustain social inequalities.[3] Moreover, the privatization and individualization of conversion as conceptualized in religious relativism do not account for the ways in which religious conversions have at times functioned to resist social inequalities.

In concurrence with Joerg Rieger's formula expounded in chapter five of this book that "materialism needs movements," I suggest that a new direction for conceptualizing religious conversion could arise from attending to religious involvement in organizing that promotes democratic participation as the means to address social challenges. The increasing involvement of religious and interreligious groups in social organizing calls for a new articulation of religious experience that takes into account the praxis of organizing. The groups involved in organizing place a priority on responding to "bodies under pressure" and transforming the political, social, and economic orders that weigh on these bodies. Throughout this chapter, I will use the term "bodies under pressure" as a new materialist frame for conceptualizing the diverse, multifaceted, and located experiences of living on the underside of social systems that compress the space bodies have to move and stretch out their lives.[4] Over the past decade, sociologists and religious ethnographers have detailed the development of religious involvement in organizing movements.[5] This scholarship has shed light on the significant role religion is playing in different organizing movements, but tends to describe organizing as a practice extrinsic to religion.[6] Based on my work with religious groups engaged in organizing, my scholarship reflects on how the religious praxis of organizing informs understandings of religious experience. In other words, how is the praxis of organizing intrinsic to religious life? For religious groups engaged in organizing, a material form of conversion that changes the orientation of bodies in society is deeply connected to religious life. We need a way of talking about the religious experience of conversion that accounts for the material form of conversion practiced in this sort of organizing. Rethinking conversion in response to the praxis of organizing promises an approach to religious conversion that respects religious difference while opening pathways for interreligious cooperation that promote social equality.

In this chapter, I rethink the religious experience of conversion in dialog with new materialisms in order to attend to the material, cultural, and political dimensions of conversion. New materialisms describe a convergence among scholars around ontologies that resist distinctions between active and passive matter. In their edited volume on new materialisms, Diana Coole and Samantha Frost describe how new materialisms move beyond earlier materialist traditions. Writing about what makes new materialisms radically new in their attention to the material world, they state:

> It means taking heed of developments in the natural sciences as well as attending to the transformations in the ways we currently produce, reproduce, and consume our material environment. It entails sensitivity to contemporary shifts in the bio- and eco-spheres, as well as to changes in global economic structures and technologies. It also demands detailed analyses of our daily interactions with material objects and the natural environment. What is at stake here is nothing less than a challenge to some of the most basic assumptions that have underpinned the modern world, including its normative sense of the human and its beliefs about human agency, but also regarding its material practices such as the ways we labor on, exploit and interact with nature.[7]

New materialisms pursue a vibrant and complex engagement of matter in light of scientific and social scientific developments that elide modern dichotomies of mind/body, energy/matter, spirit/flesh, individual/collective, and agency/passivity. As Joerg Rieger suggests in the fifth chapter of this text, new materialisms help illuminate the complexity and even the messiness of religious experience, with special attention to how power operates. The work of new materialists can help conceptualize religious conversion in a manner that includes the affective transformation of desire, the will, and the individual—but within the social, material limitations of the praxis of organizing alternative relations between bodies. Engaging with new materialisms guides my movement beyond "affective" understandings of conversion in order to articulate the religious experience of conversion in relation to the organizing of these religious groups and the complex nexuses of power concentrated in the bodies to which they respond.[8]

Approaching conversion through new materialisms resists a schism between religious conversions and material or political conversions. Instead, conceptualizing conversion with new materialisms illuminates the integral connections between material, political, and religious conversions. Specifically, I develop a movement-oriented approach to conversion as *turning, energy transformation*, and *modes of production*. Let me be clear at the outset that this description of conversion is not just metaphorical: turning, energy, and production name material processes of conversion from the atomic to the social and religious levels. I will divide my chapter into three separate engagements with new materialist understandings of conversion, beginning with Sara Ahmed, then turning to the co-authored work of Clayton Crockett, Jeffery Robbins, and Kevin Mequet, and finally ending with Jacques Rancière. I have chosen these theorists because their scholarship takes its impetus from social movements, develops non-reductive materialist ontologies, and offers insights into the material nature of conversion.

Religious Conversion as Turning: Engaging Ahmed

The recent transformations created by movements toward gender and sexual equality in religious communities have been matched by the rise of movements promoting so-called "traditional" marriage and reinforcing gender hierarchies.[9] The variety of religious movements oriented around issues of gender and sexuality are orienting (and disorienting) how bodies, identities, and communities take shape both within and beyond religious institutions. In response, I begin my reconceptualization of religious conversion by examining conversion as a form of (dis)orientation in conversation with Sara Ahmed's *Queer Phenomenology*. Ahmed's *Queer Phenomenology* is a helpful resource for expanding on affective understandings of conversion because this work represents Ahmed's own attempt to locate her earlier work on emotional intentionality within the material orientations that shape the spaces we inhabit, the relations of bodies, and toward whom and what our energies are directed.[10] In dialog with Ahmed, I contend that conversion as a form of "turning" plays a constitutive role in the (dis)orientation of bodies, identity, and community.

Ahmed's analysis of the role of "turning" in the (re)production and orientation of bodies, spaces, and communities helps us to understand the constitutive role of conversion. Ahmed argues that space is constituted through the extension of bodies into space: "Space then becomes a question of 'turning,' of direction taken, which not only allow things to appear, but also enables us to find our way through the world by situating ourselves in relation to such things."[11] Ahmed's statement is equally true at the atomic level and the social level. The very sense that things occupy space and appear visible begins with the spinning of quarks and revolving of electrons around nuclei, which mark out the surface of a space and gives the appearance that things are there. Likewise, social space is carved out through the habituation of turning bodies in this way or that way, which defines left from right, east from west, near from far, and brings certain things into view while rendering others out of view. Spaces and things are not static or given, but constituted through the process of turning.

Religious communities, then, are not static or given, but rather are constituted through conversions—patterns of turning in space that give the sense that the community is there. This understanding of conversion resonates with the sensibility that religious conversion is a foundational religious experience, constitutive of what it means to be a member of a religious community. This understanding of conversion, however, presses beyond understandings of religious conversion as punctiliar, individual conversion into a static community, in order to recognize that conversion defines an ongoing communal process. Conversion is not only what happens at the origin of a new religious community or when religious communities organize for change. The conversion of bodies continually organizes communities' identities, orientations, and boundaries—*turning* their past into their future. This does not deny that conversion may be experienced individually or punctiliarly when persons or groups get out of line with their past and head in a new direction. Rather, it exposes that the ongoing process of conversion, which keeps the community in line, tends to be obscured from view because it establishes the community's direction. In reference to the way the writing table orients the history of philosophies which have been written on it, Ahmed writes, "Despite how the table matters it often disappears from view, as an object 'from' which to think and toward which we direct our attention . . . I bring the table to 'the front' of the writing in part to show how 'what' we think 'from' is an orientation device."[12] In the same way, the past patterns of conversion of a religious community orient future conversions as the grounding for anthropological, theological, and ecclesiological possibilities.

The experience of religious conversion cannot be separated from the orientation of the broader patterns of turning that constitute the social body and may take the shape of getting in line or out of line with the inertia of the status quo. Bodies are directed to move along some paths and away from others, such that, insofar as we face the same direction as and move in step with other bodies, we are "in line."[13] Ahmed argues that "things" become oriented in space by taking certain points of view as given, which creates the sensation of lines that connect and divide different points. These lines are experienced as the sensations of surfaces, boundaries, and fixity that give space a sense of direction, relation, and differentiation.[14] Lines are not neutral: they divide bodies and space, set out ways of inhabiting the spaces they demarcate, and direct bodies in certain directions. Ahmed suggests that direction should be conceived of as socially organized patterns of turning rather than primarily a matter of individual happenstance:

> We might speak then of collective direction: of ways in which nations or other imagined communities might be "going in a certain direction," or facing the same way, such that only some things "get our attention." Becoming a member of such a community, then, might also mean following in this direction, which could be described as a political requirement that we turn some ways and not others. We follow the line that is followed by others.[15]

Space is socially organized with directional flows that constitute communities, the things to which they attend, and the things they cannot face. Religious communities are constituted through a history of conversion that gives certain spaces and arrangements of bodies a sense of sacredness, allowing communities to attend to the bodies that fit in those spaces and obscuring those that are not fit for those spaces. Insofar as religious communities get in line with the dominant social flows, they give a sense of sacredness to the inertia of the status quo. However, when religious communities turn their attention to those bodies left in the wake of dominant social flows, the conversion of sacred space reorients religious communities around "bodies under pressure."

Insofar as religious engagement in organizing orients religious communities around "bodies under pressure" that the status quo cannot face, religious conversion gives a different shape to the world and brings new possibilities (anthropological, theological, and ecclesiological) into view. Ahmed suggests that life is full of "turning points" that not only bring bodies into line in response to the address of authority, but also take us "out of line" in different directions. Turning in different directions can get one out of line, sending us on new trajectories, allowing us to face things we previously could not, and to trace other lines that mark out new worlds. She observes that, "If such turns are repeated over time, then bodies acquire the very shape of such direction. It is not, then, that bodies simply have a direction, or that they follow directions, in moving this way or that. Rather, in moving this way rather than that, and moving in this way again and again the surfaces of bodies *in turn* acquire their shape.

Bodies are 'directed' and they take the shape of this direction."[16] As discussed above, bodies in turn from the subatomic level to social level mark out surfaces that give the sense that bodies have a shape and direction. Insofar as social bodies turn in the same way, over and over again, society appears to take shape and direction. But taking alternative turning points alters a community's shape and the direction in which it is moving. Turning, then, becomes fundamental to bringing different worlds into view because it is the very turning of bodies that transforms the shape and direction of bodies, space, and communities. Organizing religious communities around "bodies under pressure" may be a disorienting experience for the community because it raises questions about the identity and future of the community. In this act of getting out of line, Ahmed writes that, "risking departure from the straight and narrow makes new futures possible, which might involve going astray, getting lost, or even becoming queer."[17] Turning out of line with authority involves a disorienting transformation of identity, space, and community. Turning, whether in line or out of line, is a conversion of the past into the future, defining the sorts of persons and things who will be at home there. Nevertheless, it is precisely the disorientation experienced in organizing around "bodies under pressure" that brings into view the past conversions of the religious community that brought it in line with the inertia of the status quo. Religious organizing around bodies under pressure gets out of line with the economic and political status quo, converting to new directions that manifest alternative ways of orienting economic space and body politics. Because the direction of bodies is not given or necessary, turning involves a performative commitment and social investment that converts history into the future. The time, energy, and resources entailed in turning and following a different line commits persons to a way of life and promises to create the sort of world in which that way of life is at home. Ahmed writes, "It is not automatic that we reproduce what we inherit, or that we always convert our inheritance into possessions. We must pay attention to *the pressure* to make such conversions."[18] If we resist the pressure to fall into line with our history by turning to attend to those things that our history could not face, we convert our history into a different future.

Understanding conversion as turning helps us to understand the perpetual nature of religious conversion that constitutes religious identities, spaces, and communities. This conceptualization of conversion resonates closely with Kimerer LaMothe's ecokinetic approach developed in the first chapter, which focuses on the patterns of movement made and remembered over time that form the relationships which create life. Conversion as turning moves beyond punctiliar and individualistic understandings of conversion by observing that, like all existence from subatomic to global levels, religious communities are formed by repeated patterns of motion that give shape to the community and determine who can be at home in the community. Even the process of maintaining a tradition requires the religious community to repeat patterns of turning that convert its past into its future. However, turning toward that which the status quo cannot face carries religious communities out of line with the inertia of tradition and the pressure to get in line with the status quo of the broader community, creating new sacred spaces and identities in which different bodies may find a home.

Religious Conversion as Energy Transformation: Engaging Crocket, Robbins, and Mequet

Whether conversion directs bodies to get in line or out of line, patterns of turning generate flows of energy. Energy conversion is central to Crocket's and Robbins's work, which is done partially in collaboration with Mequet, as they develop a new materialism in response to the global energy crisis and its ecological implications.[19] Rooted in the new understandings of energy developed in the wake of quantum theory, they claim that, "energy itself is not reductive matter but resonates with 'spirit' and 'life . . .'."[20] Recognizing that their claims exceed traditional scientific methods and enter a peculiar theological realm, they go so far as to suggest that "energy is immanent Deleuzo-Hegelian spirit (or Spirit), and energy avoids the traditional dichotomy between spirit and matter, because everything is energy transformation."[21] In conversation with the work of

Crocket, Robbins, and Mequet, this section builds on the sense of conversion as a turning that constitutes bodies, identities, and communities by developing an understanding of conversion as flows of energy that empower life.

The conversion of energy is the basis of life and the source of all (social) power. From light to plants through photosynthesis, and on to animals and humans through digestion, the conversion of energy creates life.[22] Beyond being generative of life, the conversion of energy produces power in all its forms: the power generated through physical and mental human labor, the burning of carbon-based fuels, and nuclear reactions, for instance. Likewise, the social power formed through the production, accumulation, and deployment of resources, capital, and weapons is generated through the force of energy conversion. For this reason, we should not separate our discussions of mechanical, electric, and nuclear power from our understandings of political, economic, cultural, and even religious power.

The conversion of energy generates flows of (social) power. Like all other forms of life and social institutions, religion operates in the power grid, generating flows of power or counter-power through the conversion of energy. If religion is inseparable from the flows of power, it may then be a source for alternative social power. Developing on my previous assertion that conversion constitutes bodies, identities, and communities, I suggest that conversion charges bodies, identities, and communities with flows of energy that empower certain forms of life. The history of social and global power is deeply interconnected with the harnessing and control of energy. Crocket, Robbins, and Mequet briefly trace the history of social and global power through the conversion of energy. For most of human history, social power was produced through the harnessing of primarily agricultural and animal power, and secondarily wind and waterpower. In the wake of the Industrial Revolution, however, the social and global order has been constituted through the pursuit and conversion of new "hydrocarbon dense energy" sources into economic and political power. The rapid buildup of economic and political power through the burning of carbon-based fuels that manifests in western capitalism has brought about an energy crisis. Crockett, Robbins, and Mequet contend that, "Contemporary corporate capitalism... is based upon a lie, which is the fantasy of indefinite if not infinite growth."[23] The rapid conversion of carbon-based energy into economic and political power without regard for the limited supply has created energy shortages, global warming, and massive social power differentials. Religious conversion cannot be separated from political, economic, and cultural conversions because it operates in the same power grid. The power of religion to convert communities to generate social power explains why the history of religious conversion is so closely tied to colonial expansionism: religion has often served to convert foreign bodies into sources of imperial power.

But religious conversion can also function as an alternative (social) energy source, insofar as it generates flows of counter-power that empower resistance to unsustainable modes of production that exhaust life. Crockett, Robbins, and Mequet suggest that a focus on energy conversion helps religion to become a source of power capable of reversing dominant flows of power: "In fact, with the New Materialism religion might become a source of empowerment and political mobilization."[24] They go further to suggest that, "returning to Hegel by way of Žižek, what we have is a materialist theology promising the capability to reverse the dominant power structure."[25] Because religion has played such a powerful role in the encouragement of endless energy expenditures to produce work and the authorization of environmental exhaustion, when religious movements turn their attention to the sites where the exhaustion of life continues, religious communities can generate alternative flows of power.

Generating sustainable energy sources requires more than just creating counter-flows of power against the dominant currents; resisting the currents that exhaust life necessitates organizing alternative distributions of power. Crockett, Robbins, and Mequet suggest that the energy crisis derives from the paradigm for energy conversion based on the production of heat in a closed, equilibrium system.[26] Burning fuel to generate heat has powered human expansion, but heat should not be confused with energy. While "hydrocarbon dense fuel" sources are limited, the earth is not a closed, stable system. Controlling a heat-based system is extremely inefficient and ultimately self-exhausting. Because the earth is not a closed, stable system, sustainable energy conversion depends on learning to operate in open, nonequilibrium systems with which we can coordinate, but which cannot be controlled. Addressing the energy crisis requires transforming the paradigm for power that takes the earth itself as a model of energy conversion: "The only way to achieve an energy breakthrough is to model the Earth."[27]

While Crockett, Robbins, and Mequet have specific scientific suggestions for sustainable, alternative energy sources built out of electromagnetism, their observations have consequences beyond addressing energy shortages and climate change. Their observations are equally valuable for understanding how religious movements respond to economic and political power differentials. Economic and political power differentials can only be rectified through open, nonequilibrium systems of production that are not driven through the force of social heat in a closed economic and political system. The democratic tendencies of religious movements responding to the economic and political power differentials that exhaust life redistribute power in an open, nonequilibrium organization. While such conversions of power systems make the results of religious organizing "open for debate," religious movements are capable of tapping into alternative, sustainable energy sources that empower different kinds of life. On the other hand, when religious movements operate as closed systems of social force, conversion tends to support dominant forms of power. Crockett and Robbins suggest that, "A radical theology that is freed from its allegiance to a determinate form of Christianity can dare to think not only a theological materialism but also a New Materialism of Earth beyond and behind the ideologies of globalization that are devouring us."[28] The open-ended power structure of religious movements converts the flows of energy bound up in determinate forms of religion into counter-flows capable of turning down the heat on those exhausted by the current system of production power.

For religion to attend to its material situation and become a source of counter-flows of power, it must organize open, nonequilibrium systems of converting religious energy into social power. Here, we have a way to distinguish between imperial forms of religious conversion and religious conversion that empowers resistance to hegemony. Karen Bray's development of the concept of becoming feces in the fourth chapter highlights the importance of being able to distinguish between the kinds of work energy transformation performs and the wastes it produces. Religious conversions of energy have often fueled the white-male social body through the consumption of nonnormative bodies rather than organizing religious communities around those bodies cast off as social waste. Furthermore, we have an understanding of why involvement in organizing functions to create responses to the energy crisis that exhaust communities. Religious engagement in organizing alternates the currents of power that raise the heat on those under pressure and drain bodies of life by diverting their surplus labor power. Religious engagement in organizing switches to a sustainable energy model in an open, nonequilibrium system of sharing power—by which I mean, religious involvement in organizing movements manifests a non-centralized model of decision making that respects the different needs and experiences of the diverse persons involved without predetermined outcomes.[29]

Religious Conversion as Organizing Alternative Modes of Production: Engaging Rancière

The work accomplished by alternative energies produced when religious organizing gets out of line resonates with the observations of organizing by French philosopher Jacques Rancière. During the May 1969 workers' movement in France, Rancière observed workers who were taking control of economic production for themselves.[30] They were not only defying their class-based production role during the "working day," but also transforming their class-based identities at night—by engaging in aesthetic productions such as poetry. The workers spurned capitalism's expectation that when laboring ceases, then laborers should engage in *re*storation, *re*creation, and *re*production. Rancière's observations resonate with the contemporary involvement of religious communities in economic movements in the United States. Such communities invest religious capital—the social, cultural, and economic resources that have accrued to religious communities—in giving workers a greater say in the system of production.[31] In response, this section follows Rancière's turn toward aesthetics to articulate how alternative modes of production equitably reorganize space and time. Drawing from Rancière, I contend that religious conversion organizes modes of production that give rise to desires and ideas.

Religious participation in economic organizing converts the parameters of space and time that dictate who is capable of having a say in determining systems of production. Religious participation in economic organizing does this by creating spaces and times in which workers determine modes of production for themselves. Rancière describes an aesthetic emancipation wrought by workers who defy with their bodies the space and time assigned to their class-based identity:

> In fact, social emancipation signified breaking this fit between an "occupation" and a "capacity," which entailed an incapacity to conquer a different space and a different time. It signified dismantling the labouring body adapted to the occupation of an artisan who knows that work does not wait and whose senses are adapted to this "absence of time." Emancipated workers fashioned in the here and now a different body and a different "soul" for this body—the body and soul of those who are not adapted to any specific occupation; who employ capacities for feeling and speaking, thinking and acting, that do not belong to any particular class, but which belong to anyone and everyone.[32]

In this workers' movement, Rancière witnessed the capacity of the workers to think for themselves, act for themselves, and desire for themselves in ways that defied their class-based identities. Rancière's work builds on his observation of this workers' movement to critique what he saw as philosophy's creation of a hierarchy of thinking, with philosophers self-appointed at the top. While the origin of philosophy's hierarchy of thinking extends back at least as far as Plato's philosopher king, Rancière perceived the same impulse in Althusser's work, which sought the liberation of workers.[33] Rancière challenged Althusser's underlying assumption that the role of the philosopher was to "think for the workers," with the implication that the workers' role was to carry out the philosopher's plan of liberation. The self-appointed role of the philosopher as supreme thinker elided the capacity of the workers to "think for themselves," confining workers to a class-based identity as a passive mass of followers, contradicting the reality Rancière observed in the workers' movement. Workers' movements go beyond simply changing the conditions under which workers labor so as to create an aesthetic conversion: economic organizing democratically transforms the place of workers in the system of production by organizing times and spaces in which workers have a say in the modes of production. In the case of religious participation in economic organizing, this democratic transformation changes the place of workers not only in economic modes of production, but also in religious modes of production by converting religious spaces and times around workers.[34]

The experience of conversion in religious organizing is the conversion of modes of production. Resonating with new materialist rejection of classic material dualisms, the modern aesthetic movements which Rancière terms the "aesthetic regime" inaugurate a new, nonhierarchical relationship between the visible and the say-able, modes of production and modes of thought, passive matter and active intellect.[35] The "aesthetic regime" marks a democratic trajectory in the production of art. The visibility of the arts is not subordinated to a given set of artistic techniques or an idea of what is appropriately say-able.[36] Nothing is strictly unrepresentable in the "aesthetic regime," because art is not about finding appropriate forms for acceptable subjects. Instead, the "aesthetic regime" remains indifferent to subjects of art because all things are equally capable of being the subject of art. Art in the "aesthetic regime" tends to be less about specific subjects or techniques, and more about creating certain sorts of encounters, spaces, or situations that make visible subjects who trouble the sensible logic of the day. The conversion of aesthetic modes of production that make sensible (i.e., visible and understandable) previously insensible (i.e., invisible and illogical) subjects is inseparably tied to the modern movements toward political and economic equality that give an equal say to all subjects.[37] This is not only true in terms of transforming social modes of production through economic or political activism, but also in terms of religious means of production. The involvement of religious groups in organizing democratizes who can have a say in the religious community by giving special hearing and visibility to those bodies whose sensibility is precisely unnecessary for the broader community. This is a radically democratic model of religious conversion because there is no necessary reason why "bodies under pressure" should orient the life of religious communities.[38] By organizing around "bodies under pressure," religious involvement in workers' movements makes the experience of workers sensible (visible and understandable) through conversion of economic, political, and religious modes of production.

Conversion becomes a process of feeling out new ways of being together that resist hierarchy and promote equality. Rancière's argument that modern art focuses on the creation of encounters that form new *ways of being* resonates deeply with the new materialist aesthetic epistemology of Danielle Boutet, an artist and academic who grounds her theorization in her own art practice. Reflecting on the material nature of knowledge created by aesthetic practice, Boutet writes,

> I have been led to compare creative practice to the practice of alchemy: interrogating materiality, reality and spirituality through formal and physical operations. Both the alchemist and the artist think through matter, by way of aesthetic/symbolic operations. This is not a thinking process where one finds answers to questions, but rather where one contemplates and experiences situations, themes or feeling complexes (or ways of being).[39]

Boutet's theorization of the artistic process suggests that the formation and organization of matter plays a primary role in grounding the metaphors that we think with and on which we act: "These activities [artistic processes], as forms, actually are models for ways of organizing our mind and our actions."[40] As a practicing artist, Boutet experiences the primacy of aesthetics in the generation of material ways of being that create common sense knowledge and sensible social action. Feeling out new ways of being together that give equal sensibility to "bodies under pressure" produces an open-ended process of conversion without a determinant form or goal. Addressing the ecological crises that Clayton Crockett and John

Reader raise in the next chapter in this book requires religious conversions that shift from apocalyptically oriented modes of production to ecologically oriented modes of production. Further, sustainable ecological conversions entail organizing equitable, open-ended processes for making the difficult decisions about how to transform modes of production.[41] This open-ended process of conversion reflects the experience of religious involvement in organizing that may produce specific socioeconomic commitments but the mode of production of such goals and the means to achieve them remain open-ended, because giving equal say to those previously insensible voices demands that the process not be foreclosed. The open-ended nature of such organizing movements explains why their objectives are often so slippery, sometimes leaving the public confused by their lack of clear-cut demands and sometimes frustrating those participating in the organizing process because of the slow development of concrete action. But the open-ended nature of organizing is itself the feeling out of new ways of being together grounded in equality.

Conceptualizing religious conversion in response to religious involvement in organizing reveals a form of conversion oriented to *dissensus*, which exposes the dissonance created when sensible thought and sensible forms do not align.[42] The sort of conversion experienced in religious organizing does not demand consensus, assent, or conformity to a religious content or religious form. Rather, religious involvement in organizing disorients social and religious logics and desires by orienting space and time around "bodies under pressure." The primary question at stake for Rancière in the transformation of aesthetics is whether this new division is equitable: does it present an equality of subjects? Reorganizing bodies in space and time that create alternative aesthetics renders new subjects visible and meaningful. Conversion becomes a contestation of the very divisions that organize religious and political communities and determine who gets a say and what can be seen. This form of religious conversion addresses Joerg Rieger's concern in the fifth chapter that new materialisms posit not only the earth as subject, but more specifically, oppressed persons as subjects. Engaging in alternative bodily arrangements that reapportion space and time challenges existing modes of accounting for who is a subject and what the capabilities of different bodies are. This is why the advent of the "aesthetic regime" cannot be separated from the rise of democratic politics and empirical, scientific approaches to knowledge.

New modes of production do not suspend all power relations or transcend all social divisions; rather, they divide space and time in different ways that bear the capacity for converting power relations. Whereas equality cannot be achieved, given, or won as the end of a process, equality can be the fundamental measure of the mode of production that organizes bodies in space and time and defines their subjectivity and capacities. To assume the equality of subjects converts the mode of theological and ecclesial production for religious communities by organizing around "bodies under pressure." Organizing around "bodies under pressure" opens up new ways of perceiving the divine at work in the world, and gives rise to new theo-logics and reformations of the will. Religious conversion extends from the reorganization of bodies in space and time, to include the conversion of will and belief. Organizing around new subjects gives rise to new desires and ideas, without predetermined outcomes.

Conclusion: Organizing the Future of Religious Conversion

By engaging with new materialists, this chapter has formulated a movement-oriented understanding of religious conversion that speaks to the experience of religious groups involved in organizing. Specifically, *conversion names an experience of turning that constitutes religious bodies, identities, and communities, and generates flows of religious energy that empower modes of religious production.* As I said at the outset, this description of conversion is not just metaphorical, but grounded in the materiality of turning, energy conversion, and modes of production. This framing of conversion opens new avenues for religious scholars and theologians to address the colonial history of religious conversion with attention to how some forms of religious conversion resist ongoing inequalities. I conclude my conceptualization of conversion by suggesting that rethinking religious involvement in organizing with new materialism reveals three things about the ongoing significance of discussions regarding religious conversion.

First, religious scholars and theologians must continue to discuss religious conversion because conversion organizes the identity, attention, and trajectory of religious communities. Engaging new materialism moves discussions of religious conversion beyond stultifying and individualistic understandings of conversion and toward an understanding of conversion as an ongoing process that directs religious communities. This is true not only for religious communities that are self-consciously participating in organizing, but for all religious communities. Reviving discourses about religious conversion can make us more aware of how religious communities are being directed in various ways. Such a new materialist understanding of religious conversion not only accounts for the experiences of conversion as foundational to what it means to be a part of a religious community; it also empowers religious communities to reflect on how they are organizing.

Second, the imperial history of religious conversion necessitates accounting for the political nature of religious conversion. New materialisms make such an account possible by identifying how conversion creates life as we know it, by generating flows of bio-power. Attending to these flows of power allows religious scholars and theologians to distinguish between imperial forms of religious conversion that empower political domination and forms of religious conversion that energize political resistance. We must be as clear about the capacity of religious conversion to perpetuate closed, hierarchical systems of social power as we are about the power of religious conversion to promote equitable social change. Transforming systems of political domination requires alternative sources of energy at the institutional level. Religious conversion that is oriented around "bodies under pressure" empowers such transformations. Furthermore, rethinking religious conversion with the new materialisms enables us to conceptualize conversion beyond closed, competitive forms of religious conversion, helping us perceive how conversion is operating in cooperative, open forms between religious communities.

Finally, new materialisms provide a way to discuss religious conversion beyond the will and beliefs. Discussions of conversion in terms of the will and beliefs tend to perpetuate individualistic, consumption-based approaches to change that risk not addressing the social production side of transformation. Rethinking religious conversion with new materialisms reorients the discussion to the social organization of bodies. Conversion is primarily about organizing bodies in ways that transform who gets a say in religious life and to what religious communities attend. Indeed, when bodies are organized in different ways, the conflict between experience, desire, and belief becomes evident, and alternative desires and ideas arise. Therefore, we need not separate the conversion of desires and ideas from the organizing of bodies. Instead, in the spirit of Delueze's encouragement to become a "Body without Organs," we can speak of religious conversion as "Organ-izing Bodies." Religious conversion (dis) orients religious sensibilities.

Understanding conversion within a movement-oriented framework illuminates the public, political, and material nature of religious conversion. This is not a reduction of religious experience or the unmasking of a hidden motive in religion; rather, it is a turn toward the complex material formations that religious experience takes every day. Questions of religious identity and communal organization become oriented to the ongoing, material process of organizing bodies and the manner in which this process generates power relations between bodies. The material, cultural, and political implications of religious conversion are inescapable; conceptualizing conversion in this way calls forth discourses that analyze and evaluate how different forms of religious conversion organize bodies. New discussions of religious experience in this vein are bound to be uncomfortable and messy, because they highlight how religious power is organized and to whose benefit it operates. A movement-oriented conceptualization of conversion prepares religious scholars and theologians to recognize and respond to the diversity of forms of religious conversion, with special attention to how religious conversion matters in the world.

Notes

1. Consider, for example, Joerg Rieger's treatment of soft power in the work of Bartolemé de Las Casas in *Christ and Empire* (Minneapolis, MA: Fortress Press, 2007), Chapter 4.
2. My conceptualization of religious conversion with new materialisms in this chapter follows in the vein of recent materialist approaches in religious studies, which decentralize beliefs and texts in the study of religion and emphasize the bodies and spaces of religion. See, for example Manuel Vasquez's *More than Belief: A Materialist Theory of Religion* (New York: Oxford University Press, 2010); Gregory Price Grieve's *Retheorizing Religion in Nepal* (New York: Palgrave Macmillan, 2006); S. Brent Plate's *A History of Religion in 5½ Objects: Bringing the Spiritual to Its Senses* (Boston, MA: Beacon Press, 2014). As such, my inquiry into religious conversion will not primarily focus on the changing of minds or new understandings of religious texts. Instead, my approach to conceptualizing religious conversion will be grounded in places and organizational movements. However, as I hope to show, my analysis includes a clearer sense of the role of power dynamics in the material organization of religious communities than these other materialist approaches.
3. My assertion that religious relativism is a popular but dangerous response to the violent history of imperial conversions follows recent explorations into religious difference that oppose assumptions about the unity of all religions. In this assertion, I am influenced most significantly by the work of Stephen Prothero and class-based approaches to the study of religion. Stephen Prothero's *God Is Not One: The Eight Rival Religions That Run the World* (New York: HarperOne, 2010) challenges the prominence of Huston Smith's approach to world religions which is built around the assumption that all religions are different paths up the same mountain; Prothero suggests that this well intentioned approach to religion is neither factually true nor is it useful in responding to religious violence. Prothero contends that the academic and theological claim that all religions are really the same has relativized, individualized, and depoliticized the diversity of forms that religion takes. Prothero, *God Is Not One*, 1–7. In order to deal with the material reality of religious diversity both within and between religious traditions and to establish a more helpful discourse for responding to religious violence, we need to explore how diverse religious communities make a

different impact on the world around them. However, this inquiry cannot remain purely objective because, as class-based approaches to the study of religion have demonstrated; religious difference is power-laden and the study of religion and theology must account for these power differences in order to avoid perpetuating power differentials. Examples of such class-based approaches to religion and theology include Sean McCloud, *Divine Hierarchies: Class in American Religion and Religious Studies* (Chapel Hill: University of North Carolina Press, 2007); Joerg Rieger, ed., *Religion, Theology and Class: Fresh Engagements after Long Silence* (New York: Palgrave Macmillan, 2013). Taking a new materialist approach to religious conversion creates space for a value-laden analysis of how different ways of organizing religious communities convert the world in different ways.

4. While the term "bodies under pressure" is my own, it has a genealogical history that I would like to acknowledge. I picked up the language of "under pressure" from conversations with Joerg Rieger, who has explored the idea and implications of living "under pressure" at length in his book *Grace under Pressure* (Nashville, TN: General Board of Higher Education and Ministry of the United Methodist Church, 2011). Rieger draws the language of living "under pressure" from his work with youth who would traditionally be described as underprivileged or oppressed. In collaborative attempts to describe their situation, the youth proposed the language of "under pressure" as preferable to them over terms like underprivileged. I have adopted the language as "bodies under pressure" because I find the language helpful to link the diverse metaphorical, psychological, social, and physical pressures in a way that is commiserate with a new materialist sense of bodies as ways of moving in, marking out, and expanding through space. Pressure describes the compression of space, which turns down the volume and raises the heat on bodies as they attempt to move or stretch out their area of existence. As a relative term, areas under high pressure always indicate other areas of lower pressure in which bodies can move and stretch out into space with greater volume and less resistance. Likewise, the language of "bodies under pressure" speaks to a complex dynamic of the agency/passivity of bodies that is constituted within physical, social, and cultural spaces. I hope the term "bodies under pressure" conveys a new materialist, intersectional approach to the power dynamics of systemic inequality that recognizes the conditioned agency of all bodies.

5. Consider, for example, Mark Warren, *Dry Bones Rattling: Community Building to Revitalize American Democracy* (Princeton, NJ: Princeton University Press, 2001); Richard Wood, *Faith in Action: Religion, Race, and Democratic Organizing in America* (Chicago, IL: University of Chicago Press, 2002); Jeffrey Stout, *Blessed Are the Organized: Grassroots Democracy in America* (Princeton, NJ: Princeton University Press, 2010); Helene Slessarev-Jamir, *Prophetic Activism: Progressive Religious Justice Movements in Contemporary America* (New York: New York University Press, 2011).
6. Scholarship on religious involvement in organizing has created a movement-oriented approaches to the study of religious communities, helping shift theoretical inquiries asking, "How can religion create meaningful social change?" toward material investigations into "How are religions organizing social change?" However, these inquiries tend to be one-directional, focusing on how religious communities change their broader socio-political communities. My scholarship seeks to extend this material approach to the studies of religious communities by including an investigation of "How are religious communities themselves constituted by the praxis of organizing?" This question is important because it expands the scope of the movement-oriented material approach to the study of religious communities from the ways religion changes the world around it to include the study of the nature of religion itself. Furthermore, examining the praxis of organizing as constitutive of religious communities creates space to account not only for the liberal, democratic practices of organizing that promote social change but also for conservative and hierarchical practices of organizing which might function to resist social change.
7. Diana Coole and Samantha Frost, "Introducing the New Materialisms" in *New Materialisms: Ontology, Agency, and Politics*, ed. Diana Coole and Samantha Frost (Durham, NC: Duke University Press, 2010), 3.
8. New materialisms provide the theoretical frameworks to ground affective approaches to religion that focus on the formation, experience, and conversion of desires in the material orientation of bodies. As Coole and Frost suggest, new materialisms respond to the cultural turn's failure to deal with the primacy of material limitation and the material construction of power. Coole and Frost, "Introducing the New Materialisms," 6. Most specifically, I seek to ground affective approaches to economic conversion in the material (re)organization of economic bodies. Several scholars working in the area of theology and economics have recently articulated the role of religion in the conversion of economic desires

sustaining economic inequalities, including Jung Mo Sung, *Desire, Market and Religion* (London: SCM Press, 2007); Jung Mo Sung, *The Subject, Capitalism, and Religion: Horizons of Hope in Complex Societies* (New York: Palgrave Macmillan, 2011); William Cavanaugh, *Being Consumed: Economics and Christian Desire* (Grand Rapids, MI: Eerdmans, 2008); Joerg Rieger, *No Rising Tide: Theology, Economics, and the Future* (Minneapolis, MN: Fortress Press, 2009); Daniel Bell, *The Economy of Desire: Christianity and Capitalism in a Postmodern World* (Grand Rapids, MI: Baker, 2012). While I am indebted to this recent work, I have become convinced that this understanding of conversion is not sufficient to articulate the materiality of the religious experience of conversion manifest in the praxis of organizing. Organizing is not just about affecting desires; it is more directly about converting the social arrangement of bodies.

9. Consider, for example, how religious and interreligious movements played a prominent role on both sides of the Proposition 8 debate in California, shaping the order and orientation of religious communities, social orders, and government. The identity and orientation of mainline Protestant Christianity is currently undergoing numerous conversions in response to movements for sexual equality and counter-movements opposing sexual equality.
10. Sara Ahmed, *Queer Phenomenology: Orientation, Objects, and Others*, Kindle edition (Durham, NC: Duke University Press, 2006), loc. 36.
11. Ahmed, *Queer Phenomenology*, loc. 96.
12. Ahmed, *Queer Phenomenology*, loc. 59.
13. Ahmed, *Queer Phenomenology*, loc 105.
14. Ahmed, *Queer Phenomenology*, loc. 218.
15. Ahmed, *Queer Phenomenology*, loc. 231.
16. Ahmed, *Queer Phenomenology*, loc. 241. Ahmed's understanding of turning points goes beyond Butler who contends that one becomes a subject by *turning* in response to the address of a policing power to get *in line* with its authority.
17. Ahmed, *Queer Phenomenology*, loc. 325.
18. Ahmed, *Queer Phenomenology*, loc. 270.
19. Crockett and Robbins define new materialism as "a materialism based on energy transformation." Clayton Crockett and Jeffery Robbins, *Religion, Politics, and the Earth: The New Materialism*, Kindle edition (New York: Palgrave Macmillan, 2012), loc. 135. For an analysis of religious involvement in ecological movements, see the third chapter in this volume by Clayton Crockett and John Reader.

20. Crocket and Robbins, *Religion, Politics, and the Earth*, loc. 135.
21. Crocket and Robbins, *Religion, Politics, and the Earth*, loc. 231.
22. Most fundamentally, according to Crocket, Robbins, and Mequet, "Solar energy supports life on earth." Crocket and Robbins with Kevin Mequet, *Religion, Politics, and the Earth*, loc. 1820.
23. Crocket and Robbins with Kevin Mequet, *Religion, Politics, and the Earth*, loc. 1848.
24. Crocket and Robbins, *Religion, Politics, and the Earth*, loc. 714.
25. Crocket and Robbins, *Religion, Politics, and the Earth*, loc. 729.
26. Crocket and Robbins with Kevin Mequet, *Religion, Politics, and the Earth*, loc. 2034.
27. Crocket and Robbins with Kevin Mequet, *Religion, Politics, and the Earth*, loc. 2171. Echoing my earlier reading of Ahmed's use of "turning," Crocket, Robbins, and Mequet recognize the "spin" that gives shape to life: "…everything and everyone is based upon a near or broken symmetry, a rotation and a spin that prevents simple bimorphism. This spin allows for motion, and electromagnetism drives us—subatomic particles, amino acids, humans, crystals, trees, and so on—around." Crocket and Robbins with Kevin Mequet, *Religion, Politics, and the Earth*, loc. 2195.
28. Crocket and Robbins, *Religion, Politics, and the Earth*, loc. 2994. While I understand Crockett and Robbins's concern about fixed or determinant forms of religion, their understanding of theology and religion do not reflect the genuine commitment and concreteness entailed in being part of a religious institution. For all their emphasis on the material and the constructive, there is little material attention given to religious life apart from critique of the religion of the status quo. Attending to religious groups actually involved in responding to environmental crisis would provide a helpful grounding for Crockett and Robbins's constructive discussion of religion as a source of alternative energy.
29. Consider the starfish model from Joerg Rieger and Kwok Pui-lan, *Occupy Religion: Theology of the Multitude* (Lanham, MD: Rowman & Littlefield, 2012), 121. [Editor's Note: Kwok prefers to follow the traditional Chinese pattern, wherein the family name, Kwok, is placed before her given name, Pui-lan.]
30. Joseph Tanke, *Jacques Rancière: An Introduction* (New York: Continuum, 2011), 24.
31. Contemporary religious communities have been involved in economic movements such as: Interfaith Worker Justice, Clergy and

Laity United for Economic Justice, the Workers' Rights Boards associated with Jobs with Justice, and the Occupy Wall Street movement. Economic organizing has also been a significant part of religiously supported, broad-based community organizing groups, like the IAF (Industrial Areas Foundation) and PICO (People Improving Communities through Organizing).

32. Jacques Rancière, *The Emancipated Spectator* (Brooklyn, NY: Verso, 2009), 42–43.
33. Jacques Rancière, *Althusser's Lesson*, trans. Emiliano Battista (New York: Continuum, 2011), 53–54.
34. The sixth chapter of Joerg Rieger's and Kwok Pui-lan's *Occupy Religion: Theology of the Multitude* provides numerous examples of how religious involvement in the Occupy Wall Street Movement transformed modes of religious production by converting religious space and time around workers. Rieger and Kwok write, "Just as the Occupy movement claimed the power to redefine space and create new cultural forms, the church of the multitude should also imagine creatively and not be bound by rigid traditions... In the Occupy movement, the Jewish Kol Nidre services took place in open air, instead of in a synagogue. Jews, Christians, Muslims, Buddhists and pagans shared the use of space in the faith and spirituality tents." *Occupy Religion*, 118.
35. Rancière names the new mode of artistic production that that marks the modern era the "aesthetic regime." Jacques Rancière, *Aesthetics and Its Discontent* (Malden, MA: Polity Press, 2009), 6–8; Jacques Rancière, *The Future of the Image* (Brooklyn, NY: Verso, 2007), 38–39.
36. Jacques Rancière, *The Future of the Image*, Chapter 5.
37. Rancière writes, "But the collective struggle for working-class emancipation has never been separated from a new experience of individual existence and capacities, wrested from the constraint of old bonds of community. Social emancipation was simultaneously an aesthetic emancipation, a break with the ways of feeling, seeing and saying that characterized working-class identity in the old hierarchical order." Rancière, *The Emancipated Spectator*, 35.
38. The lack of a necessity to organize around laborers simply means that workers are seen as unnecessary to the organization of society. While their bodies may produce essential goods for society, they are viewed as passive, interchangeable, and disposable. However, just because there is no necessary reason for religious communities to be organizing around "bodies under pressure," in particular, laborers, does not mean that there are not many good reasons to do so. Different religious traditions give different reasons for organizing around those bodies

deemed unnecessary to society (e.g., following a religious teaching, imitating a religious exemplar, satisfying the orders of a religious text, submission to a divine being, continuation of tradition, etc.). But what interests me here is that whatever reasons religious communities give for organizing around bodies unnecessary to society, the process of organizing around such bodies converts economic, political, and religious modes of production.

39. Danielle Boutet, "Metaphors of the Mind," in *Carnal Knowledge: Toward a "New Materialism" through the Arts*, ed. Estelle Barrett and Barbara Bolt (New York: I.B. Tauris, 2013), 30.
40. Boutet, "Metaphors of the Mind," 34.
41. Addressing issues of over-production and over-consumption of scarce resources requires new modes of production that include difficult decisions: of what should we consume less, or produce less; and who should reduce their paid labor to achieve reduced production? New, sustainable modes of production will require equitable and open-ended decision-making processes to ensure that the burden of such conversions is justly distributed.
42. Rancière suggests that the democratic trajectory of the "aesthetic regime" operates under the principle of *dissensus* rather than *consensus*. The "aesthetic regime" subverts claims to truth and to technical superiority because it breaks with the unity of form and content that underlies earlier modes of art. The unity of form and content demands *consensus* in which the sensibility of thought aligns with the sensibility of forms affirming social sensibilities and the hierarchical class structures they justify. For example, the *ethical regime* (most closely identified with Plato) prioritized sensible thought, while the *representative regime* (most closely identified with Aristotle) focused on the production of sensible forms; but both insist on the unity of form and content (Rancière, *Aesthetics and Its Discontent*, 6–8). By contrast, the "aesthetic regime" operates under the principle of *dissensus*, which exposes the dissonance created when sensible thought and sensible forms do not align (Rancière, *Aesthetics and Its Discontent*, 115). Bringing those ideas, objects, practices, and experiences that conflict with social sensibilities into view defines "art" in the *aesthetic regime*. In this manner, the "aesthetic regime" not only challenges the existing logic of "common sense" and social order, but also forges common spaces that open onto new ways of organizing society. There, previously insensible (i.e., illogical or invisible) subjects/ideas become sensible (i.e., understandable and visible).

Chapter 3

Ecology and Social Movements

New Materialism and Relational Christian Realism

Clayton Crockett and John Reader

This chapter addresses some theoretical and practical issues concerning how social and religious movements might address urgent environmental concerns. Specifically, we advocate a confluence between what has been called the New Materialism and a more specifically Christian movement, Relational Christian Realism. Bringing these two terms together, we suggest, offers fresh insight into some of the most important ecological and political questions we confront. Furthermore, this chapter is the result of a collaboration by two scholars and theologians, one in the United Kingdom and one in the United States, who share viewpoints and perspectives across both countries with implications for other parts of the world.

What are the possibilities for the relationship between theology and environmental concerns? First, we must consider the possibility that there is no inherent relationship, as they are different discourses working on the basis of completely different assumptions or world views. The relationship created between them will then depend upon which discourse or concern is the dominant or driving force, and what outcomes are desired. If the aim is to remain true to a particular interpretation of Christian doctrine, then environmental considerations will be secondary. If, on the other hand, the objective is to encourage particular ethical behavior that is consistent with environmental objectives, then those objectives will be paramount and doctrinal issues become of secondary importance.

Most writing in the field falls somewhere between these two extremes, but ends up blurring the relationship as a result. So some doctrines are given attention whereas others are deemed superfluous or unhelpful, or else some environmental concerns are given priority but others receive little attention as they are harder to square with familiar doctrinal positions. If it were the case that certain interpretations of Christian doctrines or practices have led in a linear fashion to appropriate environmental activity, then it becomes hard to explain the failures over time of Christians to engage with these issues. If it is understood that any such relationships are contingent and haphazard, then it becomes less obvious why Christian positions should be important at all, as they are unlikely, in themselves, to lead to the appropriate environmental activity.

Instead of a direct and seemingly arbitrary relationship between Christian theology and environmental concerns, one could propose combining Christian insights with those from other discourses or traditions, in which case the relationships become even more complex and difficult to negotiate. Who decides which elements of which disciplines are to be combined or brought into relationship? According to which—and whose—criteria are these relationships to be negotiated? Where is the balance to be struck between the ideas and theories of the different parties? If the driving force is to motivate or encourage appropriate ethical responses to a situation described and justified by nontheological sources—for example, the science of climate change—then why expect theological reasons to appeal to nonbelievers or to be of any use or relevance in the debate? In this case, one is only trying to "preach to the already converted" and to get them to take environmental concerns seriously by linking these to familiar or acceptable Christian ideas. Do we really believe that Christian ideas actually have something to offer to the process that other disciplines lack in some way? If not, what exactly does Christianity "bring to the party" at all?

Taking all this into account, we argue that a new discourse is required to interpret the developing context and to articulate appropriate practical responses. The question then is: how and where precisely are Christian faith and practices to play a part in this, and how are they to be related to the concepts and language developed from nontheological sources? One might talk of this process as being blurred or entangled, but then this itself rests upon a specific ontological position that assumes an acceptable level of inter-relatedness and which requires both nontheological and theological warrants. If that is the starting point, then traditional theological doctrines will play a subsidiary or secondary role unless it can be shown that they add something fundamental to the mix.

An example of how this works out in practice can be found in two recent books that focus specifically on the issue of climate change: Michael Northcott's *A Political Theology of Climate Change* and his co-edited book with Peter M. Scott, *Systematic Theology and Climate Change: Ecumenical Perspectives.*[1] The latter book in particular struggles with the question of how to balance environmental concerns with traditional theology, and while each chapter begins with a brief overview of how and why climate change is an issue for theology, each then goes on to examine the familiar areas of Christian doctrine—the Trinity, Sin and Salvation, Eschatology, and so on—in an attempt to construct a coherent response from within these fields. The exception is chapter 6 by Rachel Muers on "Creatures," which at least creates a wider scope for investigation. One might question why the current focus on is climate change as if it were the only, or the most important, environmental issue, although it might be because discourse on climate change is closest to being a metanarrative and thus can be placed in the same format as traditional theology, which makes similar metanarrative claims for itself. The problem, however, is that it is not easy to square a metanarrative that throws the whole future of humanity on the planet into question, with the idea of a positive and determinate outcome or *telos* from within traditional interpretations of the doctrine of Creation. There is no consensus among the diverse authors of the book on this issue.

What we propose in this chapter is an alternative approach and discourse which moves away from the idea of any specific metanarrative, either theological or environmental, and concentrates instead on a wider range of issues, dealing with them at a micro- or meso-level rather than exclusively at the macro-level. Such an approach is able to address the complexity of areas, balancing the detailed and empirical with the more theoretical and conceptual. This approach has been more recently developed in two interrelated trajectories: the New Materialism and Relational Christian Realism. After describing these entangled approaches, we will examine some specific examples of environmental concerns and responses, showing how the theoretical and practical are combined at this level of focused analysis.

The New Materialism is a way to name the affirmation of material reality in a complex manner, in contrast with a crude, reductionist atomic materialism. The New Materialism emerged as a term during the late 1990s, primarily in connection with interpretations of the philosophy of Gilles Deleuze. In works like *Difference and Repetition* and his collaboration with Félix Guattari on *Capitalism and Schizophrenia* (*Anti-Oedipus* and *A Thousand Plateaus*), Deleuze elaborates a differential ontology that resonates with contemporary mathematics, ecological systems theory, chaos, and complexity physics. These readings of Deleuze stress an alternative materialism to classical modern materialism. The New Materialism affirms the significance of diverse processes that occur on distinct plateaus—energy transformation, cellular reproduction, biological populations, animal consciousness, human self-consciousness, and language—but refuses any hierarchical scale of value. This is the meaning of Deleuze's phrase "plane of immanence." Not that everything is reduced to the same status, but the affirmation of the plane of immanence constitutes a refusal of transcendent meaning given from elsewhere. The meaning or significance of an event is immanently constituted, and events happen everywhere.

Proponents of this New Materialism include Rosi Braidotti, Jane Bennett, Manuel De Landa, and William Connolly. Other proponents of the New Materialism look for inspiration from Alfred North Whitehead's philosophy of process, sometimes in connection with Deleuze, like Isabelle Stengers, or go back to the philosophy of

Maurice Merleau-Ponty, like Diana Coole. Often an implicit or more explicit feminism is affirmed by the New Materialism, informed by the philosophy of Luce Irigaray. These various trajectories within New Materialism coalesced during the late 1990s and flourished in the first decade of the twenty-first century as a theoretical movement, and we affirm its continuing importance for philosophy, ecology, and even theology.

More recently, religious and theological readers have taken up and engaged with the New Materialism. Many religious and theological perspectives oppose materialism, as it seems to reduce the spiritual and transcendent to physical, material things. The New Materialism, however, is non-reductionist because it is based on levels of energy transformation. Energy is a material reality, but it can be conceived as a spiritual reality at the same time. In their book *Religion, Politics, and the Earth*, Crockett (one of the co-authors of this chapter) and Jeffrey W. Robbins sketch out a post-capitalist vision based largely on the New Materialism, where being is energy transformation, and the Earth is viewed as subject rather than object.[2] Here is a radical theological materialism that opposes any dualism between spirit and matter, or transcendence and immanence. In terms of theoretical religious studies, in his book *More Than Belief*, Manuel Vasquez develops a non-reductive materialist theory of religion that "highlights complexity, inter-level connectivity, emergence, situated knowledge, and relative interdeterminacy and openness against monocausal, unidirectional, and totalizing schemes."[3]

In addition to its insistence on the significance of material reality, another reason to affirm the term materialism is its political charge. The history of revolutionary political thought is indebted to Karl Marx, and Marx's dialectical materialism is key to his understanding of history. Marx stands Hegel back on his feet, and views history as the development of material economic processes such as the interaction of labor and capital. We do not embrace Marxism as a dogmatic truth, but we refuse to vilify or abandon Marx's insights. In a post-Marxist context, the New Materialism affirms an energetic cosmology and ecological ontology, but it also retains a critical, political edge. Religion is the opium of the masses, but it is also a force for social transformation, and theology is not simply about escaping the world, but it is also a force for changing it.

All things are processes, systems, or objects—we do not draw sharp lines between these terms. Things work based on transformation of energy, which is infinite and always conserved while at the same time entropic, or tending toward a limit in practical usage. Our political and environmental reality is ultimately an energy crisis, and we are using up available finite natural resources, at least in amounts that could be renewable on the scale of human lifetimes. As we reach these energy thresholds, we desperately strive to become more efficient even as the overall activity of human production grows to an unsustainable level and reaches a tipping point. Factory farming of large animals for food is one instance of this general problem, as we will see below; global warming is another. We are currently undergoing what Elizabeth Kolbert and others call The Sixth Extinction, and it is not clear that we can avoid becoming victims ourselves.[4]

Theology concerns not only the energy and activity of God, but the workings of human and other beings with which they are intimately interrelated. Energy is the source of life and value, although it manifests in different forms and processes. Energy is always in a state of transformation; it is never static. Theology is about what is most significant or important, our highest value or ultimate concern. Usually this ultimate concern is called God but, for many contemporaries, faith in a transcendent God has become questionable. As our situation becomes desperate, we may hope and pray for a god who could save us, as Heidegger says in a posthumously published interview.[5] If only a God can save us and yet there is no transcendent God, then we are lost, because God will not save us, and we cannot save ourselves. But if we are not ourselves, because "we" are different at every instant, then there is nothing to save, and this is good news, because we are always in transformation.

The New Materialism is more than just an ontology or cosmology—it is also a way to live politically and socially. Another, more recent and more explicitly religious movement is Relational Christian Realism (RCR). RCR, as elaborated in the United Kingdom by Chris Baker, Tom James, and John Reader in their book *A Philosophy of Christian Materialism: Entangled Fidelities and the Public Good*, is a reaction on the theological level against the theological idealism and imperialism of Radical Orthodoxy, and is an attempt to reconnect with the realism of Reinhold Niebuhr and the more philosophically inflected approach of Paul Tillich.[6] It would be a mistake, however, to see this theological realism as the sole thrust of this new movement. RCR has many parallels with the New Materialism in terms of the sources upon which it draws, and it extends what has previously been understood as realism well beyond familiar boundaries, including links between philosophy of religion and religious practices.[7]

Whereas the realism of Niebuhr and the philosophical theology of Tillich focused upon the purely human dimensions of creation, RCR is in tune with the concerns for the nonhuman and thus able to respond to the environmental issues, as will be shown in this chapter. It also raises the question of whether it is legitimate to propose any definitive telos or purpose in creation itself because RCR acknowledges the essential entanglements of the human with the unknown and unpredictable futures of nature as a whole. The relational aspect central to RCR sees the human as always already part of the nonhuman in its broadest sense, challenging any notion of human exceptionalism. It builds instead upon the flat or immanent ontology more closely associated with the New Materialism.

In terms of method, its main characteristics are as follows. Rather than assuming that any specific religious tradition can be self-sufficient, it draws upon ideas and insights from other disciplines and other traditions, thus exercising a humility that acknowledges the limits of the human capacity to interpret and respond. RCR looks critically at all public debates, especially those that claim to be based upon a neutral standpoint, and argues that since all such debates always already presuppose certain values the task is to articulate these debates so that alternatives can then be brought into the discussion. RCR adopts the approach developed by Bruno Latour, which presses for a methodology that reassembles and takes things slowly rather than preempting debates and decisions by the introduction of grand summaries or essentialist categories. Specifically, Latour attacks the modern division between nature and politics, arguing that we need to envision a widened conception of social actors that compose a common social assemblage or college.[8] Here, one needs to attend to "matters of concern" that possess the opportunity to surprise us rather than would objective "matters of fact."[9] All of this is undertaken with the understanding that this is about process rather than end product, and that, as Latour himself affirms, the task is to keep the references circulating.

Drawing also upon Deleuze, RCR examines material religious practices by identifying the rhizomatic nature of practical activity rather than focusing solely on doctrines and beliefs. For Deleuze and Guattari, a rhizome is a way to imagine an extensive flat multiplicity without hierarchy or unity: "we will therefore speak of a plane of consistency of multiplicities, even though the dimensions of this plane 'increase' with the number of connections that are made on it."[10] Here, religious experience is performative as much as discursive, linking to other spheres of human activity such as politics and environmental concerns but without assuming that we are capable of forming definitive or closed answers to the realities we now face. Attention to the assemblages or machines in which we are entangled and implicated is here seen as a more appropriate ethical response than one which relies upon some notion of human autonomy or a theological imperialism that assumes it already knows the end point of creation as embodied in the church as idealized institution.

On a political level, RCR advocates the need to act at both the micro- and macro- dimensions, but recognizes that change and challenge are more likely to occur by operating in the spaces or interstices which escape institutional control. Examples given in the book *A Philosophy of Christian Materialism* include the reforms of the education system in the United Kingdom, proposals to allow fracking to access new sources of energy, and the progressive localism that sees alternative political practices emerging in spite of centralized government control of welfare and community work. In all of these situations, the task is to work out what it means to be faithful, and to continue to work out how to live out the values of faith traditions in the midst of the entanglements and compromises that are inescapable for those who are involved in direct activity.

Linking once again to some of the ideas of the New Materialism, the most controversial aspect of RCR for many traditional Christians is probably its adoption of the work of Deleuze, Levi Bryant, Braidotti, and Latour, specifically, by embracing immanence rather than transcendence as the most appropriate basis for a Christian ontology. The reasons for this are both philosophical and political, and these new conceptions and practices challenge the structures of hierarchy and power that are associated with transcendence in its familiar forms.

As with the older concepts of theological realism, RCR resists the notion that there can be a singular alternative polity or specific economics that can be associated with the Christian tradition, but it nevertheless argues for alternatives to current configurations of capitalism by searching for nomadic, pragmatic and complex entanglements with the real. There is thus an openness to both what humans might become (nobody knows what a body—or machine—can do) or what wider nature (including human culture) may yet emerge from current possibilities. As we have discussed, the human is always already and inescapably implicated and entangled with the nonhuman, and it is only out of this continuing inter-relationship that the future will evolve. Humans are no more and no less than pilgrims across the plane of immanence, finding and losing their way as they seek to embody the virtual, which is the space of consistency or rhizomatic multiplicity. Here the virtual is not opposed to the actual, but complements it. These ideas have implications for other areas of more traditional Christian doctrine, which are explored in A Philosophy of Christian Materialism, but it is the consequences for environmental concerns that are of direct interest to this chapter.

Having described and interrelated these two approaches—the New Materialism and RCR—we can now turn to engage a couple of books that stand broadly within the same general trajectory. Inadequate or questionable conceptual frameworks can be correlated to damaging and harmful practices, as they both shape and are shaped by the material consequences of our actions. Therefore, in the opening section above, we raised the issue of how to bring different discourses into a real encounter, challenging the assumption of clear and distinct demarcation lines between, in this case, theology and environmental concerns. Then we described the New Materialism and Relational Christian Realism as complementary movements that are well placed to address these interlinking concerns between environmentalism and theology.

Having already described the difficulty and necessity of entangling discourses and fidelities, it is helpful to read in Whitney A. Bauman's recent book, *Religion and Ecology: Developing a Planetary Ethic* that he too argues that the notion of pure discourse is itself a result of the forms of thought and practice associated with late capitalism. He asks, for instance, what religious tradition is not always a hybrid of every tradition with which it has come into contact. So what is Easter without Passover, the German tradition of the Easter bunny, or the Roman traditions celebrating the spring equinox? Bauman says that

> such isolated commodification/reification of identities, families, traditions and products is precisely what capitalism depends on. These theological and metaphysical traditions help cocreate the world and universe as a Newtonian collection of objects. In other words, these substance-based ways of thinking help to create the very reified, isolated worlds of late technocapitalism.[11]

So the ways in which discourses are presented and then related, or not related, are never neutral, but always already represent a certain set of values that have political and practical consequences.

Having established this relation, the next question concerns how a particular religious self-understanding needs to be challenged on the grounds of its connections to that capitalist worldview. A doctrine of creation that proposes an essentially linear view of time and an end point or telos toward which creation is inexorably moving, can be seen as implicated in this critique, as it is based on certain metaphysical assumptions that are shared with capitalist modernity. Bauman says that all such stories are successful because they hide their own mechanisms of production and maintenance and appear instead as simply "natural." Instead, he offers some alternative and disruptive stories based on a notion of radical immanence. As one finds in Spinoza, the counter-idea to telos is that nature does not progress in any preordained direction, but becomes and develops in many different possible directions and with no essential natures. "There is no 'natured' but only the continual process of nature becoming."[12] So, rather than some grand metaphysical claims about the future that might provide a foundation from which one can evaluate the present, we would accept the openness and indeterminacy of the future and therefore the possibility of unforeseen transformations: "no one knows what a body or nature can do." Within such an alternative account, there is therefore also the possibility of other ways of understanding the relationship between the human and the nonhuman.

Once one has identified in the discourse that something is natural, this conclusion presupposes that other things are not natural. Similarly, once one draws a clear boundary around what is supposedly human, the rest is identified simply as nonhuman. Bauman, along with Latour, argues that environmentalists often assume that nature is all that is not human, but the consequence of this is that the "other," or the abject (that which is excluded from the process and rejected), that which is not human, cultural and technological, is denied as part of the overall process of becoming. Yet, it is these objects that are essential to the processes. Such abject others or objects need to be redefined and repositioned if environmental issues are to be adequately addressed. This is what Bauman criticizes as "human exceptionalism," and he concludes that any definition of nature which does not include humans, imagination, religions, cultures, technology, and ideas will inevitably fail. In this he agrees with Latour that:

> The point of recognizing the abject is not that somehow there will be a transcendent point at which all life is included, but rather to continuously trouble the permeable edges of our thought toward ever new ways of becoming that seek to bring the claims of the abject to our attention. Such a focus on process helps us to come to terms with the fact that we are always already coming into being in a collective process of repetition of differences.[13]

As those already familiar with the work of Whitehead or Deleuze and Guattari will recognize, Bauman is moving here toward their conceptual terminology of process, rhizome, and flat multiplicity along with an immanentist approach that brings into question any metaphysical or foundational understanding of nature, human, or nonhuman, and indeed, questions the boundaries that are often drawn between them, even by environmentalists. He suggests that both the latter—and indeed anti-environmentalists—are often afraid of the same thing: change. Changes and developments in technology, as well as changes in everyday habits, become the subject of a more general fear of change, threatening either a future that is running out of control, or one that challenges and disrupts a supposedly comfortable and sustainable status quo. Both fail to acknowledge that "life is an emergent process that is open, uncertain, and always a bit messy."[14] There is no bottom to nature, nor a top for that matter, but rather the multiple levels or folds (to use Deleuzian language) from which one can analyze or experience nature at all its different levels, including the physical, chemical, biological, social, cultural, and religious.

This approach takes us a long way toward challenging the subject-object and human/nonhuman distinctions, distinctions that are challenged by the flat ontologies associated with both the New Materialism and Relational Christian Realism. It can also link these views with environmental concerns such as those raised by another recent book, written by Philip Lymbery, called *Farmaggedon: The True Cost of Cheap Meat*. This book does not specifically advocate a religious or theological perspective, but it does bring together many of the ecological, social, and political concerns that we need to address, and of course the title, Farmageddon, invokes a religious and apocalyptic subtext for the dire content. Farmageddon shows how the intensification of factory farming of animal products contributes to the perpetuation of human and animal suffering, as well as environmental devastation across the planet. As the consolidation and intensification of factory farming grows, mainly for production and consumption of chickens, pigs, and cows, one can detail the massive costs in the form of pollution, the food and water necessary to raise these animals, and the health risks that result from so many animals being raised and killed in such confined spaces.

The book itself is pessimistic, even though the authors strive to point out the benefits of non-factory farming, including financial ones, and sketch out alternatives in the form of organic and humane practices of raising and eating animals. The basic framework, however, is the depletion of natural resources, which conflicts with the fundamental economic model of capitalism, which is premised on the indefinite growth of resources. As human population and activity grows in size and impact, other species, the environment, and the atmosphere are negatively affected, and there is, furthermore, a feedback loop that impacts human beings as well.

So let us examine in greater detail two examples of the questions being raised by Farmageddon. The intention is to identify the conceptual framework that underlies our approaches to farming and to suggest how a different focus—that upon immanence rather than transcendence—might be able to transform both theory and practice. The first example is the UK Dairy Industry, which has been transformed by commercial and technological impacts over recent decades. For instance, 30 years ago, the average UK dairy cow was producing 5,000 liters of milk per year, but the figure now is closer to 7,000 liters. This requires that they be fed high-energy concentrated food in order to sustain them and to counteract the pressures of this greater intensity of production. Some herds produce closer to 10,000 liters a year, and this means that grass cannot provide for their food requirements and has to be replaced by grain.[15] This dietary change makes it pointless to turn cattle out to grass, and instead they have to be kept indoors for much of their lives. There are already mega-dairies being built in the United Kingdom to cater to this alternative approach. Within this system, the role of the vet is to keep animals healthy for long enough to be productive, and it seems they are unlikely to raise issues of animal cruelty since they are themselves fully integrated into this system. The system also benefits feed, equipment, and pharmaceutical companies, who achieve economies of scale with these larger enterprises.

Another result of this transformation is that many smaller producers have gone out of business in recent years, as they cannot compete with the larger producers. Though there have been high profile campaigns in the United Kingdom to protect milk prices and such smaller producers, the economic imperatives work against this proposed solution. The link with the major supermarkets has also come under scrutiny, as the supermarkets may be able to determine and dictate prices to the farming community because of their near monopoly of milk sales. Large milk processing firms have also increased their hold on the industry and are in a position to shape the future. Having been to a state of the art processing plant, one notes that the product could be almost anything, as this is simply another industrial process, often staffed by workers from other EU countries who are more desperate for work at lower rates of pay. Does this all matter? Perhaps this is the inevitable direction of change once commercial factors become predominant and technical innovations contribute to ever faster and more efficient processing of the product, but there are impacts upon humans and nonhumans alike that simply are swept aside in any public debate, as it feels as if a juggernaut rolls its way through the dairy industry.

In the book itself, Lymbery refrains from arguing that "small is beautiful," but he could probably go further than he does in raising questions about the underlying assumptions that sustain these current practices. Once so much is invested in what becomes an industrial process, it is difficult to challenge this process on the grounds of impacts upon both animals and human communities, as the power rests with the large companies who can dictate terms to others and for whom milk production is no more than of instrumental value. If the consumer wants the cheapest product at greatest convenience, then there will come a point when overseas production replaces domestic. Polish milk has already been imported into the United Kingdom, bringing into question the future viability of the UK dairy industry. Once the small scale and local dairy units have disappeared, they are gone forever, as it is not financially viable to reestablish them.

One can see similar factors at work in the second example, that of the pig industry. The book recounts the story of a beach in Brittany that, as a result of pollution from the growth of pig units, has become unsafe for tourists and locals, and so hotels are struggling to keep going.[16] Lethal algae, commonly known as sea lettuce, have taken over, spurred on by an excess of nitrogen carried downstream by polluted rivers and waterways. As the algae dry up and decay, they release hydrogen sulfide, and since 2008, this gas has apparently led to the death of wild animals in the area and to other environmental damage. The problem of course, is that this is big business. Brittany produces 14 million pigs a year, and the region is France's top pork producer, with over half the country's pigs being reared in this area. However, one is hard pressed to spot any pigs out in the open as this is industrialized rearing, driven by modern machinery, cheap labor, and biotechnology.

Feeding this pig population is a major enterprise in its own right, requiring cereal crops that are then processed into animal feed, plus soya, which is sourced from around the world, often from deforested land in Brazil. At the other end of the process, as it were, the pig waste is gathered, processed into liquid manure, which is used as agricultural fertilizer and spread on corn plantations. Polluted waterways are the result, which then leads to other environmental problems and is also a health hazard for humans. Alternative and more sustainable methods do exist, where animals can be fed mainly from what is grown on the individual farms and they live happily on straw beds, but those implementing this approach are seen as mavericks.[17]

Similar issues are found but writ even larger in the United States. According to the book, the commercial hog industry in North Carolina is huge and worth $2 billion to the economy. At any one time, there are around 10 million pigs being raised on factory farms in the state. It is estimated that in one North Carolina county alone, 2.2 million pigs generate as much untreated manure as central New York City creates sewage. All of this waste has to go somewhere. In 1995, there was a catastrophic spill when a lagoon filled with waste ruptured and 117 million liters poured into the headwaters of the New River. This spill was twice the size of the Exxon Valdez oil spill 6 years earlier. Although environmental legislation has improved, there are still huge lagoons of pig sewage carrying a variety of unpleasant matter. Disposing of the levels of waste produced by factory farming is a major issue, and a scientist working for one environmental pressure group has estimated that the total cost of cleaning up after the US pig and dairy industries could approach $4.1 billion.[18]

These are just a few examples of the areas of concern raised by the book, which lead to troubling conclusions about how farming across the globe is creating more problems than it is solving. The impact of this study is to make one question meat consumption and production and to wonder how we got ourselves into this situation in the first place. Beyond what has been described, other problems include: the role of technology and use of antibiotics and terminator seeds; wastage of food itself and the political importance of this; crops being grown to feed factory animals not humans; the poor nutritional value of animals and birds bred this way; cruelty to animals and birds; high rates of suicide among Indian farmers trapped by agricultural programs; the regular issue of power and how it is being exercised in order to sustain and protect the existing farming regimes. So what comes out of the book is far more than the obvious concern for the welfare of animals, but a range of issues related as much to the wellbeing of humans and communities. The clear failing of conceptual thought is that of not seeing the interrelatedness of human and nonhuman and presupposing what might be called a human exceptionalism, based upon a hierarchical or metaphysical superiority of the human. As we have argued, the necessary conceptual challenge to this humanism comes from an understanding of immanence derived from the New Materialism and Relational Christian Realism.

This further step can be illustrated by reference to Levi Bryant's book *Onto-Cartography: An Ontology of Machines and Media*, where he develops what has been called an Object-Oriented Ontology (OOO). A focus on objects is one way to counter the implicit humanism of philosophy's single-minded focus on subjective consciousness and language in the twentieth century. OOO, as developed by Graham Harman and others associated with the movement of Speculative Realism, counters this subjectivism and also allows us to view objects in much more complex and nuanced ways.[19] In his book, Bryant commends Ian Bogost's idea of an alien or nonhuman phenomenology, and suggests what this might mean in terms of the relationship between humans and cows.[20] The objective, for Bryant, is to switch perspectives and to see the world from the point of view of the nonhuman, in this case, the cow's, in a way that is not simply the subjective adoption of another's perspective but a more radical transformation of identity. We are challenged to distinguish our aims from those of the other entities, objects, or "machines" in

Bryant's terminology. By attending to how cows experience the process of slaughter, for instance, campaigners have been able to adapt the practices in ways that are less traumatic. This may seem like a minor response, but as part of a larger change of perspective, it might contribute to changes in material practice that are less harmful to all involved. It's not just a simple lessening of trauma to make animals' lives slightly less brutal; it's an entire transformation of thinking about what it means for these objects to be interacting in these ways. This transformation requires that humans begin to recognize the extent to which our beliefs and practices are shaped by those other "machines" that form our common life and experience. Such a perspective leads us beyond what Noëlle Vahanian calls "organic thinking" in her critique of the dairy industry, in the direction of a more "machinic thinking" that involves the "institutionalization of the imagination."[21]

This very different view of the world can thus have implications for the ways in which we approach the whole range of environmental and indeed political issues, and it has strong theological consequences. In an ontologically flat world, there can be no sovereign beings, or even hierarchy of beings with humans at the top: "there is no machine that is beyond being affected by other machines."[22] Therefore, if a God exists, he must be a tinkerer like the rest of us and must contend with the exigencies of other machines. An ecology is a network of machines without a single governing principle; hence, all ecologies are anarchic. The contributions of both Bauman and Bryant coming from a New Materialist perspective begin to show how an immanentist rather than metaphysical approach challenges the traditional boundaries between subject and object, human and nonhuman, and has important implications for both environmental practice and religious self-understandings.

Conclusion

How can these theories of New Materialism and Relational Christian Materialism, which are already entangled with each other in this chapter, be put into practice? First, we must recognize that thinking is already a material practice, one that takes place among "machines" or assemblages of brains, body, social pressures, and habitual practices, all along a plane of immanence. Second, we must redirect our energies, which are social, psychic, religious, political, ontological, and theological, in more productive orientations.

Transformational events happen along the edge of chaos rather than in a steady state. Transformation is about seeing and reinforcing the linkages among distinct ideas, people, and practices, pressing them toward progressive social and political change while recognizing how difficult if not impossible such a task may be. Many religious organizations are caught up in questions of dogma and doctrine, arguing over beliefs as well as the orthodoxy social-moral practices. We refuse orthodoxy in any form except, perhaps, the weak notion of praise of the rightness or goodness of creation, at least its potential goodness. Evangelical Christianity summons powerful energies and redirects them toward practices that support and sustain what William Connolly calls the "evangelical-capitalist resonance machine."[23]

In contrast, we seek to liberate these energies for transformative social and political practices, partly by helping readers view energy, theology, and matter differently. One practical example here would be the Transition Towns movement that began in Totnes, England in 2004, based on permaculture practices as a response to the peaking of global oil production, largely associated with the ideas of Rob Hopkins.[24] This proliferation of transition towns has transformed into a Transition Network of cities, towns, and localities across the world. These transition initiatives stress the interconnection of all aspects of our lives and work to shape and reshape a more vital and sustainable practice using less energy and producing less waste. Although these transition communities are not explicitly religious, they cultivate a new form of spiritual discipline that works against contemporary global capitalism.

Finally, an example of Relational Christian Realism at work in response to the range of issues raised by Farmageddon is to be found in the United Kingdom at a retreat house in Worcestershire on the banks of the river Avon, not far from the Birmingham and West Midlands conurbation. With an explicit Christian foundation and a commitment to introduce and practice environmentally friendly policies, this retreat center is attempting to live out a lifestyle that acknowledges the need for a greater harmony between nature and culture. Among other things, it aims to work the garden and grounds organically, not using any pest controls or fertilizers; produce as much of its own fruit and vegetables as possible; compost any waste food; install a beehive to produce its own honey; and has constructed a Bible Garden trail and natural history garden in order to raise awareness among those visiting and staying at the house. All food is freshly prepared and cooked on the premises, and meals are designed to minimize waste by not producing excessive helpings. Meat-based meals have been reduced so there are two days of meat, two of fish, and three vegetarian. Food is locally sourced wherever possible and the fish supplier has been selected on the basis of being able to trace a sustainable and appropriately bred source. Local beers and fruit juices are used, and Fair Trade products are also available. All paper and printing materials are recycled. In other words, the aim is to set an example and to show that it is possible to conduct these normal activities in ways that do less damage to the environment and encourage a greater sense of relatedness between the human and the nonhuman.

This is not to suggest that it is possible to live out an idealized lifestyle where there are no ambiguities or compromises, but rather, to recognize the entangled, incarnated, and fully implicated dimensions of human activity, as we struggle to respond to the risks and dangers created by the ways in which we produce food, aware of the pressures of land use and population growth. Such retreat houses in the United Kingdom find it difficult to be financially viable even without this sort of stricter regime and many who use it may not be fully in favor of the principles behind it, but this is one way of troubling the permeable edges of our thought. In contrast to the United States, where many evangelicals still resist environmentalism, it is largely the evangelical wing of the British churches that have taken up environmental causes through groups such as the John Ray Initiative based at the University of Gloucester, A Rocha UK, and A Rocha International, and their respective links with the Christian Rural and Environmental Studies Course which is based at Ripon College, Cuddesdon near Oxford, the biggest Anglican ministerial training college and nonresidential course. In the United States, some people and groups comprising the Emerging Church movement have pushed for a deeper connection between ecological and religious practices and perspectives, and a growing number of Christian evangelicals support the Evangelical Climate Initiative, which began in 2006 and takes seriously the threat of global warming.

Although many of these environment and social movements would not share the more radical theological and philosophical viewpoints articulated in this chapter, they are certainly in the vanguard in terms of trying to practice an environmentally sensitive form of Christianity, engaging in both local projects and publication and training. What is still lacking, though, is a more appropriate conceptual framework, and it is here that we argue that the ideas of the New Materialism and Relational Christian Realism could add a crucial dimension to the growing engagement between ecological concerns as a New Social Movement and faith-based entanglements.

Notes

1. See Michael S. Northcott, *A Political Theology of Climate Change* (Grand Rapids, MI: Eerdmans, 2013); Michael S. Northcott and Peter Scott, eds., *Systematic Theology and Climate Change: Ecumenical Perspectives* (London: Routledge, 2014).
2. See Clayton Crockett and Jeffrey W. Robbins, *Religion, Politics, and the Earth: The New Materialism* (New York: Palgrave Macmillan, 2012).
3. Manuel A. Vasquez, *More than Belief: A Materialist Theory of Religion* (Oxford: Oxford University Press, 2011), 5.
4. Elizabeth Kolbert, *The Sixth Extinction: An Unnatural History* (New York: Picador, 2014).
5. See Martin Heidegger, "Only a God Can Save Us," trans. Maria Alter and John D. Caputo, *Philosophy Today* 20 (1976): 267–284.
6. Chris Baker, Thomas A. James, and John Reader, *A Philosophy of Christian Materialism: Entangled Fidelities and the Public Good* (London: Ashgate, 2015).

7. See the Philosophy and Religious Practices Network based at the Universities of Liverpool and Liverpool Hope in conjunction with the William Temple Foundation, University of Chester, UK. http://philosophyreligion.wordpress.com/.
8. Bruno Latour, *Politics of Nature: How to Bring the Sciences into Democracy*, trans. Catherine Porter (Cambridge, MA: Harvard University Press, 2004), 18.
9. Latour, *Politics of Nature*, 79.
10. Gilles Deleuze and Felix Guattari, *A Thousand Plateaus: Capitalism and Schizophrenia*, trans. Brian Massumi (Minneapolis: University of Minnesota Press, 1987), 9.
11. Whitney A. Bauman, *Religion and Ecology: Developing a Planetary Ethic* (New York: Columbia University Press, 2014), 103.
12. Bauman, *Religion and Ecology*, 38.
13. Bauman, *Religion and Ecology*, 59.
14. Bauman, *Religion and Ecology*, 134.
15. Philip Lymbery with Isabel Oakeshott, *Farmageddon: The True Cost of Cheap Meat* (London: Bloomsbury, 2014), 126.
16. Lymbery and Oakeshott, *Farmageddon*, 176.
17. Lymbery and Oakeshott, *Farmageddon*, 181.
18. Lymbery and Oakeshott, *Farmageddon*, 186.
19. See Graham Harman, *The Quadruple Object* (Hant, UK: Zero Books, 2011).
20. Levi Bryant, *Onto-Cartography: An Ontology of Machines and Media* (Edinburgh: Edinburgh University Press, 2014), 65.
21. Noëlle Vahanian, "Great Explanation," in *Religion and Violence in a Secular World: Toward a New Political Theology*, ed. Clayton Crockett (Charlottesville: University of Virginia Press, 2006), 164, 168.
22. Vahanian, "Great Explanation," 115.
23. William Connolly, *Capitalism and Christianity, American Style* (Durham, NC: Duke University Press, 2008), Chapter 2.
24. See Rob Hopkins, *The Transition Handbook: From Oil Dependency to Local Resilience* (White River Junction, VT: Chelsea Green Publishing, 2008).

Chapter 4

Becoming Feces

New Materialism and the Deep Solidarity in Feeling Like Shit

Karen Bray

"Pure energy is being itself, which is an absolute materialism, materialism degree zero. But energy plus one, plus two, and so on is material complexity, a folding of being on itself. It is becoming thought, becoming time-image, becoming a brain."[1] These words help to conclude the chapter "Being (a brain)" in the recent work by Clayton Crockett and Jeffrey Robbins, *Religion, Politics and the Earth: The New Materialism*. In this, their ontology chapter, Crockett and Robbins draw on Catherine Malabou's concept of neuro-plasticity. As Crockett and Robbins note, for Malabou, "plasticity indicates a form that possesses three characteristics: the ability to give form, the capacity to receive form, and most importantly, an explosive plasticity, the charge of an auto destructive aspect of form itself."[2] Since the brain is an always in-process, plastic entity, Crockett and Robbins's "Being a Brain" is unlike Hegel's assertion that "Being is Thought," in that it attempts to break open the dualisms between matter and thought and conscious and nonconscious. The metaphor of brain, then, for Crockett and Robbins, disentangles determinism, stasis, and immateriality from a sense of being that is tied to thought. "Becoming brain" marks an eventive being—a being capable of creativity *and* destruction.

This New Materialist ontology provides great opportunities for postmodern thinkers eager to return to the very material matters of embodied living. And yet, I cannot shake a sense of unease with the call to become like a brain. My dis-ease is not over the call to understand becoming as energy transformation that reflects neuro-plasticity. This ontology feels phenomenologically sound and politically potent. Additionally, inspired by the concept of plasticity, Crockett and Robbins propose: "a radical theology that is freed from its allegiance to a determinate form of Christianity [which] can dare to think not only a theological materialism but also a New Materialism of Earth beyond and behind the ideologies of globalization that are devouring us."[3] And yet, is this metaphor fully breaking open the dualisms that Crockett and Robbins hoped it would? Could this non-dualistic neuro-philosophy unintentionally risk an anthropocentrism? What if it problematically overlooks raced and gendered concepts of "brain"? Might Crockett and Robbins—while working to find metaphorical counters to the deterministic ontologies and ideologies of globalization that are devouring us—have missed an opportunity to address more forcefully those of us whose bodies and identities have been disproportionately devoured, and devoured on the premise of *not* being like a brain; those of us whose blood and sweat have been extracted, consumed, and used up by those in power—only to be discarded as the waste, the shit of history?

Further, is the call to become like what has already been prized enough to disrupt deterministic anti-earth theologies, to challenge narratives of divine—or human, or male—mastery and superiority? Or might a call to be a brain continue to risk a drift upward—a more material drift, to be sure; an energetic drift, but one, still, that moves inward to the self and upward to the head, that master over all of the body's messier movements?

According to New Materialist Mel Chen, to what we grant animacy has powerful implications for systemic oppressions built around race, gender, sexuality, and ability.[4] Hence, to reaffirm the animacy of the "brain" may not redress all of the ways in which we have degraded and deprioritized "inanimate" material traditionally divorced from consciousness. Additionally, in centering our ontologies on the animacy of the brain, Crockett and Robbins might be unintentionally skirting the ways in which we have made objects of those human and nonhuman animal bodies considered anything but "brain."

So where might we find a linguistic, material, and metaphorical supplement to an ontology of becoming brain—a supplement which equally evokes a sense of energy transformation and plasticity, yet one which cannot materially be detached from our other-than-human entanglements, and which cannot linguistically be detached from those of us who have been marked as body, object, beast, flesh—essentially, those of us who in our assumed lack of brain (i.e., rational thought, mind) have been expelled from, and through our expulsion propped up, the white male social body? Perhaps we might find it in the shit or feces, in the energy of our digestive tracts. Yes, we should be becoming brain, with all its form-giving and form-taking capacities, but are we not also becoming shit, with all its impurities, material entanglements, smells, and discomforts, its taking of forms, but also its giving form to new life, to new possibilities, and connections beyond the human body?

In search of our fecal ontologies, I turn to Gayatri Spivak's classic review of *Glas*, Jacques Derrida's typographically unorthodox work in which the column on the left engages Hegel, that philosopher of mind, and the one on the right plays with the "fart artist," Jean Genet. According to Spivak, "*Glas*...tries to undo the hierarchy between body and mind, inviting us to oscillate between Hegel and Genet, the classic classifier and the masturbating fart-artist, and see how far they are accomplices through a necessary self-betrayal."[5] Spivak refers to the Genet column also as the fecal column. I find in this column a metaphor for being—one that (Glas so helpfully reminds us) always runs alongside and aids in self-betrayal with mind. Indeed (perhaps ironically) we might read the Derrida of *Glas* as an early New Materialist. His very deployment of *Glas* and its various linguistic and definitional inclinations reveals a return to the potency of the material body and its messy entanglements:

> Why "gl"? Because, at least in one reckoning, the unrestrained economy of *Glas* goes from the dictionary meaning—"knell," the sepulchral tolling by which the corpse is declared to be not mere body and thus given significance—to the sound of the spit in the throat and the death rattle; in fact, to all the sticky agglutinative gurgling of the body as such. Reading with Derrida involves a perpetual undoing of the intelligibility-corporeality hierarchy that this range conceals.[6]

This undoing of the intelligibility-corporeality hierarchy achieved by the throaty range of "gl" reflects an undoing hoped for by Crockett and Robbins, but one—I argue—that might need a more sticky metaphor than brain in order to succeed. For how might we honor the knowledge produced from the standpoints of those of us who have often been thought of as body, beast, and object—those of us whom have not just been becoming, but were already assumed to be waste? This essay proposes the sticky supplement feces as an apt addition to a hybrid and radically democratic metaphor for being: becoming brain/becoming feces.

What exactly might this hybrid metaphor do? Or rather how might it move? How might it move us? How might becoming brain/becoming feces move the social body, animating new approaches to social movements and political theologies? How might a return to the potency of that which we think of as disposable help us to rethink a religio-ethical call to be ever more in communion with all of creation? What kind of God might live in the shit? What kind of religious practices—prayers, laments, praise, liturgies, worship, and study—might honor such a God? Theologically *and* politically, what might it mean to be becoming brain/becoming feces? Placing Joerg Rieger and Kwok Pui-lan's concept of "deep solidarity" into dialog with queer and feminist theories of New Materialism and affect, this essay argues that if those of us in the 99 percent are going to adequately find deep theological and political solidarity with all of its members (most importantly, those made most precarious and considered most disposable by neoliberal capitalism), we will need to experience the discomforts and contingencies inherent in those who have been considered disposable and worthy of expulsion from the social as well as the religious body. I suggest that if we are to more intentionally challenge the historical contexts of power and domination that have created hierarchies (not only between mind and body, but also between the organic and the inorganic), we will need to get into the muck, to look to the materiality that questions our own sense of mastery, success, and ability. If we are energy transformation then we might (like the plastic brain) become able to compose new narratives of the self; but also we might unbecome, like excrement, allowing our sense of self-enclosure, determinism, and wholeness to decompose.

The Animacy and Agency of Shit

In *Vibrant Matter*, Jane Bennett argues that when we begin to expand our understanding of where subjectivity, agency, and vibrancy can be found, to include nonhuman and inorganic matter, onto-theological binaries are challenged and a new brand of politics that better recognizes the influence of nonhuman actants is nurtured. Indeed, to challenge the very ideologies that Crockett and Robbins worry are devouring us, Bennett looks to that which we are literally devouring. In her chapter on "Edible Matter," she discusses how food has agency. Within a given assemblage, an edible actant can behave similarly to what Michel Serres calls a "thermal exciter," which can, "make the equilibrium energetic distribution fluctuate."[7] In other words, it can cause energy transformation. Food has long been acknowledged as the source of calories, of energetic units necessary for life. Still, Bennett's concept of vibrant matter helps us to recognize how our very reliance on what we consume can destabilize any sense of self-determination. Indeed, Bennett notes that "edibles disclose... what Deleuze and Guattari called a certain 'vagabond' quality to materiality."[8] What Crockett and Robbins find in the plastic quality of the brain, Bennett locates in the nonhuman edible. This understanding of the revelatory powers of edible matter fortifies the appropriateness of the metaphor of becoming feces. Our materiality and ontology are transformed through what we consume, how we digest it (or fail to do so), and—after it becomes a part of us—how we discard it/us through the joint endeavor of becoming feces.

Additionally, Bennett argues that food is not simply a tool to be used up by the acting human, but rather is an actant in assemblages not only with our digestive tracts, but perhaps also with our moral sensibilities. For instance, contemplating how any number of vibrant material bodies (human and otherwise) might form agential assemblages, Bennett asks, "Did the typical American diet play any role in engendering the widespread susceptibility to the propaganda leading up to the invasion of Iraq? Do sand storms make a difference to the spread of so-called sectarian violence?... Can a hurricane bring down a president? Can HIV mobilize homophobia or an evangelical revival? Can an avian virus jump from birds to humans and create havoc for systems of health care and international trade and travel?"[9] By asking these ontological and political questions, Bennett helps us to see the importance of a metaphor for being that takes seriously the other-than-human actors in our agential capabilities. In light of Bennett's propositions, "becoming brain/becoming feces" could be read as the acknowledgment that our very becoming in the world is reliant on a number of vibrant materials, including those microbes that live in our bellies and those edibles that make their way through our digestive tracts. Further, the addition of "becoming feces," might helpfully check the hubris engendered when agency is granted only to "conscious" actors.

While Bennett's work is crucial for challenging the possible anthropocentrism found in the metaphor of brain, it is through engaging the work of Mel Chen that the gendered and raced implications of "becoming brain" can best be brought to light. According to Chen, linguists have denoted what they call an "animacy hierarchy, which conceptually arranges human life, disabled life, animal life, plant life, and forms of nonliving material in orders of value and priority."[10] For Chen, animacy hierarchies in Western ontologies are about kinds: different kinds are assigned varying degrees of generativity—some kinds are marked as reproductive and others as nonproductive.[11] Linguistically, we might say that "brain" marks a particularly generative kind. While they might prefer the term "eventiveness" rather than "generativity," it is this sense of the plastic brain's potentiality (even through destruction) that excites Crockett and Robbins. It is the very fact that the "brain" is of the kind that can create and take on form that makes it a particularly apt metaphor for being as becoming, as energy transformation. Linguistically, therefore, "brain" as a generative kind attaches to those society deems vital: whites, heterosexuals, cisgendered men, the wealthy, the able-bodied, and, significantly for theology, a begetting and creating Father God. Conversely, "shit" or "feces" can linguistically be understood as a nonproductive kind, the very part of us that does not serve a generative purpose for the continuation of our beings, as it is the part we expel. Like the "gl" of the sticky agglutinative gurgling of the body, shit sticks to those deemed socially disposable: people of color, immigrants, indigenous peoples, sex workers, the homeless, women, queers, transgender folk, the disabled, the mentally ill, the poor, the nonhuman, and ultimately (and catastrophically) the Earth itself. Yet, of course, feces are incredibly generative. Without organic waste, nothing new can grow. The gap between the linguistic animacy and the actual animacy of feces is precisely the abyss from which a new theological and ontological construction might be found. For instance, Chen notes that animacy hierarchies are themselves plastic. They can slip and give, yet they do not do so divorced from their ties to privileged positions of race, gender, ability, and sexuality. Kinds associated with the assumed nonproductivity of shit might include the disabled body, the animal body, and the queer body—all bodies that have been assumed to be social waste.

What might happen if we were forced to recognize the animacy of shit and so the animacy of "kinds" linguistically associated with the nonproductivity of fecal matter? Might recognizing how we have both linguistically and socially ejected these bodies from the social body—from a narrative of generativity, and so purposefulness—help us to rethink what types of productivity, eventiveness, and success we have prized? Might we rethink what types of divine relation and religious practice we have valued and pursued?

Feces can be rethought as a site from which we find new ways of thinking our very beings and our energetic responsibilities to become along with all other actants in our assemblages. We might rethink the importance of the theological insights, the mattering, of those bodies too often made inanimate and considered degenerative. As Chen argues, "It seems that animacy and its affects are mediated not by whether you are a couch, a piece of metal, a human child, or an animal, but by how holistically you are interpreted and how dynamic you are perceived to be."[12] An ontology centered on "becoming brain" risks perceiving dynamism where it has always been found, within the human, the man, the mind. However, if we are to find dynamism in the shit, to perceive its holism, to see it for all its life giving capabilities—its ability in its very acts of decomposition to allow for new composition—might we also learn how to never stop seeking to interpret one another more holistically? In order to practice a fuller recognition of one another we will need both to understand alternate forms of generativity *and* to be willing to risk finding solidarity in the feces, striving not only to become brain (with both its plastic potential and its problematic associations), but also to get into the muck that comes with feeling like shit.

The Fecal Ontology of Occupy

In response to the critiques that paint the Occupy Wall Street movement as primarily white and middle class, Joerg Rieger and Kwok Pui-lan argue that this middle class involvement in Occupy is "a more important accomplishment than it might seem at first sight, especially when the middle class begins to realize that it is not out there merely to help others but that it is itself deeply affected by the economic crisis and thus shares in some structures of oppression with the rest of the 99 percent. This is the first step toward what we are calling 'deep solidarity.'"[13] From a liberationist point of view, the call for deep solidarity, as explicated in Rieger and Kwok's recent collaboration, *Occupy Religion: Theology of the Multitude*, is a call not to be on *the side of* the oppressed, but rather to recognize how one is similarly, if not identically, oppressed. For Rieger and Kwok, a false sense of superiority or distance between those traditionally considered middle or upper middle class and those considered working class prevents the kind of solidarity needed to address the tragic realities created and nurtured by neoliberal capitalism:

> In the past we members of the American middle class have often felt that we were benefiting from the wealth and power of the 1 percent. Even if we have not consciously reflected on it, we have considered ourselves to be in closer proximity to the ruling class than to the working class. Many of us have intuitively agreed with the often-repeated but never proven mantra of neoliberal capitalism that a rising tide will lift all boats... But the reality is that the fate of the middle class resembles more and more the fate of the working class: our savings and stockholdings are getting smaller and smaller, our jobs are less secure, and our influence is declining.[14]

Deep solidarity in this way resonates with my call to become feces. In order to cultivate a deep solidarity, those in the middle class cannot just make the traditional liberal political move of siding but not identifying with the working class. Rather, we all must try to understand and, I argue, *feel* the growing precarity and disposability to which we *all* are increasingly subjected. Recognizing our own precarity, however, will not be enough. We must also recognize the privileges to which we still cling—our reluctance to truly unbecome in an act of deep solidarity.

"Becoming shit" is a call to give up the comforts[15] (including the *false* comforts of prestige that come through participation in neoliberal, white supremacist, and heteronormative systems that may not be killing our souls at the same pace or intensity, but which are indeed toxic for us all) sought out by kinds associated with reproductivity. This would include the form of productivity associated with what Rieger has named the "Rising Tide Theology." This rising-tide theology carries with it bold faith claims: that economic deregulations promote growth, that tax cuts for powerful corporations and the wealthy spur the economy, and that wealth gathered at the top eventually trickles down.[16] These claims rely on the partisan and never-innocent belief in a providential market. The belief in a rising tide is reflective of what affect theorist Lauren Berlant has diagnosed as "cruel optimism." According to Berlant:

> A relation of cruel optimism exists when something you desire is actually an obstacle to your flourishing. It might involve food, or a kind of love; it might be a fantasy of the good life, or a political project. It might rest on something simpler, too, like a new habit that promises to induce in you an improved way of being. These kinds of optimistic relation are not inherently cruel. They become cruel only when the object that draws your attachment actively impedes the aim that brought you to it initially.[17]

To believe that our fortunes will rise as those at the top rise is a cruelly optimistic belief, as the rising tide of wealth causes increasing levels of misfortune for all those left in its wake. Hence, those of us made disposable by the rising tide are left to blame ourselves, told we have not produced enough, succeeded enough, been enough; we are led to believe we signed our own death warrants.

Berlant has additionally traced the affective resonances projected by neoliberalism that shape our relations to self, other, and the world. For instance, in her chapter, "Slow Death (Obesity, Sovereignty, Lateral Agency)," Berlant argues that slow death involves the slowly wearing away of bodies through what is commonly thought of as the obesity epidemic in the United States. Unlike the neoliberal assumption that the diets of poor people come from ignorance or laziness, Berlant diagnoses a matrix of influences that contribute to this epidemic: "Apartheid-like structures from zoning to shaming are wielded against these populations who come to represent embodied liabilities to social prosperity of one sort or another. Health itself can then be seen as a side effect of successful normativity, and people's desires and fantasies are solicited to line up with that pleasant condition."[18] Perhaps most interesting here is the rhetoric of longevity and futurism that promises happiness within a period that may never be achieved by bodies subject to this slow death. Berlant argues that this experience of a curtailed future combined with the feeling of quotidian crisis and exhaustion often results in the search for momentary pleasure and satisfaction in food. Hence, the optimism found in the promise of pleasure from the meal becomes cruel in that it shortens the life span of the diner. This is the kind of cruel futurism that was offered up by the subprime mortgage boom and subsequent crisis, and is nurtured through the predatory practices of payday lenders. It is a futurism of cruel optimism that can be countered by a different kind of hope—one found through the rejection of successful normativity.

This hope through rejection often resides in subcultural living and politics—for instance, those found throughout the queer community and, recently, the Occupy movement. Many in the queer community who have resisted successful normativity have done so having already been marked as failures, as the waste. For instance, Jack Halberstam has argued that we should "leave success and achievement to the Republicans, to the corporate managers of the world, to the winners of realityTV shows, to married couples, to SUV drivers."[19] For Halberstam, queer artists who create worldscapes in resistance to neoliberal concepts of success become politically potent failures. To be a failure in this way is, "with Walter Benjamin, to recognize that 'empathy with the victor invariably benefits the rulers.'"[20] Empathy with the victors is exactly what Rieger and Kwok are critiquing when they note the middle class problem of identifying with the 1 percent. Becoming a failure then is finding deep solidarity and is another name for becoming feces. To become feces is to refuse to identify with those who seek out mastery: mastery over one's own body, mastery over the social body, and mastery over the body politic. Hence, an ontology of becoming feces is the opening to a theology counter to narratives of divine sovereignty that measure piety through acts of obedience to and imitation of an Imperial Master God. To embrace failure, shit, and deep solidarity is to find holiness not in the seat of the throne, but in the spirit that moves among those in the streets, the classrooms, the beds, the wheelchairs, the tent cities, the libraries, the slums, the fields, and in the event that is and was the Occupy movement.

As reflected in Rieger and Kwok's "deep solidarity," there is significant overlap in the cultural politics of failure described by Halberstam and those embodied in the solidarity found within Occupy. Occupy's fracturing of decorum and public/private boundaries and the discomfort willingly embraced by occupiers are reflections of a Halberstamian political failure, one that says we will not play a game that has been rigged against us. In this way Occupiers began, and continue, to find deep solidarity with those historically considered failures: the queer, the African-American, the woman, the child, and the global poor.[21] Occupy is a movement that asks us to become both brain and shit. It looked and looks to actants divorced from traditional liberal politics. It resisted and resists the cruel optimism of Wall Street in favor of forming an alternate public square.

Both in her October 23, 2011 visit to the General Assembly of Occupy Wall Street held at Washington Square Park and in her plenary address at the 2011 annual meeting of the American Academy of Religion (AAR), Judith Butler spoke of just this kind of body politic. Due to the banning of megaphones, Occupy rallies communicate through the "people's microphone," in which the crowd repeats, and so magnifies, what is said. Hence, even when star speakers like Butler address the crowd, those gathered serve as the microphone repeating her words in unison. Butler's voice, as such, became simultaneously amplified—creating something new through what we might call an act of becoming brain—and diffused and dependent on the democratic crowd assembled (becoming feces in solidarity with the anonymous masses). For instance, when she proclaimed, "I am Judith Butler," so did dozens of others. When she proclaimed, "As bodies we need each other in dependency and desire. So this is a politics of the public body," so did dozens of others.[22] The people's microphone allows *us all* to become the expert theorists, while we are also all becoming the public body. Further, to understand the politics of Occupy as a politics of the public body reflected in these very conditions of desire and dependency is to begin to feel what it might be to be becoming feces.

Shit is a metaphor of being that can help us materially grasp the kind of precarity enforced on the 99 percent and, in particular, on those made most vulnerable by the financial structures that undergird the growing disenfranchisement of American and global populations. For instance, during her speech at the AAR, Butler, drawing on Berlant's "slow death," mentioned that there was "a heightened sense of disposability differentially distributed in society."[23] This differential of course exists between the 1 percent and the 99 percent, but so too is disposability differentially distributed within the 99 percent. Realizing this differential is a key to deep solidarity. Further attunement to this differential might lead to a greater willingness to become brain *and* feces, and to the lessening of our resistance to the very messy entanglements of vibrant material reflected in becoming shit. To be attuned to the differential is to work to *feel* the material reality of what it is like to be considered disposable. For instance, as Butler noted in her speech, society has deemed disposable those who cannot afford healthcare and those conscripted into the army and sent into conflict zones where there is no clear mandate: their lives are forfeitable. To explicate this point, she retold the story of the September 12, 2011 GOP presidential debate aired on CNN. During the debate, when moderator Wolf Blitzer asked Ron Paul whether an uninsured 30-year-old workingman in a coma should be treated, Paul responded that the man should assume responsibility for himself. When Blitzer followed up by asking whether the congressman was saying that society should let the man die, a joyous cheer of "Yeah!" rippled through the crowd. This celebrated death sentence inflicted on the uninsured by Paul, this proclamation of disposability, was met with whoops of affirmation. From this anecdote, Butler asked us to consider "under what conditions joyous forms of cruelty arise."[24] To become like shit is not just to recognize the problem with this joy, the cruelty in the adulation, but is also to live into the despair that accompanies the infliction of such cruelty. It is to be affected by the materiality of this emotion, and by the precarity of those it condemns as disposable, as waste. For, as Butler—again drawing on Berlant—noted, "precarity is a structure of affect."[25] Just as theologian James Cone has argued that one is either on the side of the lynchers or those lynched,[26] we can proclaim that one must either be on the side of those made joyous by this cruelty or those made depressed, melancholic, or enraged. One either empathizes with the victor or finds deep solidarity with the "losers." Hence, we may have to disrupt more than just our political and class standings. In order to live into the materiality of deep solidarity we will need to disrupt our privileged emotional flows in the world. This disruption might come in the form of micro tactics of the self, or in new styles of worship and prayer, or in theological and political ruptures instead of redemptions; but no matter its form, its affects will involve discomfort. For, to ignore the affectual dimension of challenging one's privilege might be to practice a cheap solidarity in place of a deep one.

Feeling like Shit

We cannot adequately address a New Materialist approach to the problems of neoliberalism, hierarchies of kinds implied in (re)productivism, and cheap solidarity without addressing the materiality of affect and emotion. For Chen, the turn toward animate materiality cannot be divorced from the study of affect, the way certain material objects and subjects imply certain affects or feelings such as happiness or despair.

In *Animacies* Chen illuminates the deep racial and sexual implications that come both from a greater sense of vibrant materiality and from "negative" or "positive" feelings. Chen situates her work at the intersection of each of these fields: "*Animacies* draws upon recent debates about sexuality, race, environment, *and* affect to consider how matter that is considered insensate, immobile, deathly, or otherwise 'wrong' animates cultural life in important ways."[27] This understanding of race and sexuality has vital consequences in terms of a New Materialist ethic for social movements and theological reformations. For instance, Chen argues that inorganic matter has often served in its own vibrant way to solidify the "dangerous" or "deathly" character of certain raced and sexed bodies. Chen explicates the racializing role of matter in her chapter on the recent scare over lead toxicity in Chinese toys. According to Chen, lead took on a racialized and sexualized personification as the predator from without that had somehow infiltrated our homes. For instance, she notes that representative headlines during this scare read: "'Lead pulls the trigger,' and, 'The Enemy at Home.'"[28] By making lead the subject, these headlines reveal a sense of animacy in that which we ordinarily think of as inanimate. Lead revealed its agential character, and one configured as nonneutral. According to Chen, lead's threat was conceived as such because its toxicity was targeting *white* children; it was a foreign invader of a white domestic sphere. The resultant hysteria, Chen notes, did not arise as significantly when it came to poor black children living in apartments in inner cities where lead paint was chipping away and slowly poisoning their families. Nor did this panic acknowledge what the effects of lead's toxicity might mean for laborers in China who were producing the toys. As Chen puts it, "The image of the vulnerable white child is relentlessly promoted over and against an enduring and blatant background (that is, unacknowledged) condition of labor and of racism: the ongoing exposure of immigrants and people of color to risk that sets them up for conditions of bodily work and residence that dramatize the body burdens that projects of white nationalism can hardly refuse to perceive."[29]

The differential between the media coverage and societal sensibility surrounding the poisoning of black children and Chinese laborers and that around the poisoning of white children reveals a differential between the disposability of people of color and whites. We might here say, then, that to be in deep solidarity with black children is to be on the side of disposability. To be in deep solidarity with the disposable is to attempt to feel what it is like to be permitted to be poisoned. To become feces is to ask what it would take for us to put our bodies in the line of toxic fire. It is to consider for whom we would give up our gas masks. It is to face our unwillingness to be exposed in such a way, and to ask what this unwillingness means for those children whose very homes have been killing them. It is to ceaselessly ask these questions and work to be and become differently until those considered more disposable than us are no longer in poison's sight. These questions arising from the play on how and to whom and to what we grant subjectivity are key to ethical propositions sprouting from queer, affect, and New Materialist theories and, hence, are crucial for an understanding of their theological and political potential. Particularly important for a reformulation of solidarity à la Rieger and Kwok is the tie between Chen's New Materialism and a queer focus on a politics of negative feelings or affects.

According to the editors of *The Affect Theory Reader*:

> Affect arises in the midst of *in-between-ness*: in the capacities to act and be acted upon . . . Affect is found in those intensities that pass body to body (human, nonhuman, part-body, and otherwise), in those resonances that circulate about, between, and sometimes stick to bodies and worlds, *and* in the very passages or variations between these intensities and resonances themselves.[30]

Affect theory might be considered the critical exploration both of what types of acts, knowledge, bodies, and worlds are produced in this in-between space *and* of how we might better think affect's role in such a production.

By taking into account the material phenomenon of affect in the world, and the production of socially mediated feelings, affect theory reaffirms the significance of the dependent and desirous character of the material body. For instance, in *The Promise of Happiness*, Sara Ahmed examines the affects and effects of "happiness," and how the objects it forms shape our ontological and political relationships with society. Through her critical reading of "happiness," Ahmed identifies those who, through their distance from socially mediated "happy objects," are considered failures or threats. These threats, or what Ahmed calls "affect aliens," include the feminist killjoy, the queer, the revolutionary, and the melancholic migrant. Affect aliens are those who do not fit the affectual script handed down by mainstream society. For instance, she notes, that, "to be a good subject is to be perceived as a happiness-cause, as making others happy. To be bad is thus to be a killjoy. This book is an attempt to give the killjoy back her voice and to speak from recognition of how it feels to inhabit that place."[31] The killjoy inhabits a space of negativity and threat. From within this space of threat Ahmed finds possibility. This possibility rests in what can be learned from pausing awhile and inhabiting the terrain tread by affect aliens. It is by walking the paths of the feminist killjoy or the revolutionary that we might better practice deep solidarity.

For Ahmed, feelings are pedagogical. The promise of happiness teaches us what to associate with happiness. And as such it teaches us to desire to be associated with such associations. Through socially mediated emotions we are taught value, and through the learned experience of what society deems properly valuable we are engendered to be either satisfied or dissatisfied. In the epistemological and pedagogical force of feeling, Ahmed finds a sense of political hope in the ways in which we might reorient happiness: "we might need to rewrite happiness by considering how it feels to be stressed by the very forms of life that enable some bodies to flow into space."[32] For instance, in her discussion of Alfonso Cuarón's film "Children of Men," based on the book by P. D. James, she notes that the revolutionary character does not flow easily. The revolutionary is stressed, experiencing the world from the position of resistance. Ahmed asks what we might learn from this experience of counter flow, and suggests that, "we might revolt by revolting against the demand for happiness."[33] To revolt against the demand for happiness is not to reject the utility of happiness. Indeed, Ahmed finds happiness interesting. It is, rather, the rejection of the idea that to be happy is to be free. Ahmed suggests that true political freedom comes when we are free to be unhappy, particularly in a society that engenders such unhappiness: "The freedom to be unhappy would be the freedom to live a life that deviates from the paths of happiness wherever that deviation takes us. It would thus mean the freedom to cause unhappiness by acts of deviation."[34] In this reformulation of political freedom, *The Promise of Happiness* aims to tap into bad feelings as creative responses to an unfinished history.[35]

It is in the risk of feeling unhappy, the risk of deviating from the affectual norm that I believe those of us traditionally at the center, traditionally considered to be becoming brain and not shit, can cultivate deep solidarity. The risk of feeling unhappy takes on a religious character in theologies attuned to trauma and injustice. These theologies refuse an easy cure, a God of peace and quiet, and a too quick resurrection. For instance, in her constructive pneumatology, *Spirit and Trauma: A Theology of Remaining*, Shelly Rambo challenges Christian theologies and church practices that seek to rush from Good Friday to Easter Sunday, passing too quickly over the day that remains in the wake of the traumas of crucifixion—Holy Saturday. Rambo asks what remains after death, after trauma, or, as we might put it after having been made disposable: "What does life mean in the moments when one leaves the graveyard, walks home, and crawls around and cannot see her way forward? Meaning is dead. Hope is dead. Love is dead."[36] Drawing on theologies of witness she finds from within both the Holy Saturday theology of Hans Urs Von Balthasar and Adrienne Von Speyr and the teachings of John's Gospel, Rambo helps us to rethink how love remains embodied in a middle Spirit: "The Spirit provides a distinctive way of orienting oneself between death and life, a way of witnessing the fractured dimensions of word and body between death and life."[37] This orientation is nurtured—I argue—in micro and macro tactics of becoming feces along with brain. To become feces in this way is to witness to what remains in the wake of crucifixions, when all meaning and hope seem to have been lost. To become feces, to be an affect alien, is to become an impediment to the too quick push to resurrection. It is a an act of faith that chooses to feel the pains of neoliberalism, white supremacy, heteropatriarchy, and ableism at the expense of one's own assured resurrection. It is to be in deep solidarity with what and who remains after and in the midst of trauma. This deep solidarity will mean not merely *listening* to those that have been violently expelled from the social body, but also working to *feel* the pathos of their unhappiness, their melancholy, their rage, and deciding to become rageful with them—no matter how discomforted we are by such intensity of feeling. Fecal acts of faith will be those that block the affectual flows that push us too quickly to Sunday, to impede those unwilling to feel the pain of Friday on Saturday. A theology of the affect alien—the feminist killjoy, the shit—calls us to hear the laments of those on the emotional frontlines.

In response to the tragic August 9, 2014 shooting of Michael Brown, an unarmed black teenager, by a white police officer in Ferguson, MO, bloggers began posting Lauryn Hill's song, "Black Rage." It is sung to the tune of *The Sound of Music's* "My Favorite Things" and includes the following lines:

> Black rage is founded: who fed us self-hatred
> Lies and abuse while we waited and waited?
> Spiritual treason, this grid and its cages.

Victims of violence both psyche and body
Life out of context is living ungodly.
Politics, politics
Greed falsely called wealth
Black rage is founded on denying of self!
Try if you must but you can't have my soul
Black rage is founded on ungodly control![38]

Black rage as expressed in this song embodies the political freedom to be unhappy. It is also a call—a lament—that seeks out a politics and theology of affect, one with which we might only find deep solidarity if we are willing to feel more deeply: to rage, to weep, to get politically depressed.[39]

Affect theorist Ann Cvetkovich argues that "[in] a utopia that doesn't make a simple distinction between good and bad feelings or assumes that good politics can only emerge from good feelings, feeling bad might, in fact, be the ground of transformation."[40] Hill's song does not elicit a feel-good politics; nor does "Hands up Don't Shoot," the slogan used by protesters across the country in the wake both of the Brown shooting and the consequent militarized police occupation of Ferguson; nor does #YesAllwomen, the Twitter hash tag that went viral in response to the response to the mass shootings committed by Elliot Rodger in Isla Vista, California on May 23, 2014. In videos released following the shootings Rodger proclaimed deeply misogynistic feelings, feelings that seemed to be in part the rationale behind his violence. Social media and the blogosphere were then set aflame with articles about the misogynistic and rape culture that women face on a quotidian basis. In response, a Twitter hash tag #NotAllMen went viral. #NotAllMen was a way to claim that not all men were bad, murderers, abusers, rapists, or misogynists. What #YesAllWomen said back was that not all men are rapist, but yes all women have been scared they will be raped; not all men make threatening advances, but yes all women have been made to live under threat; not all men are actively participating in misogynistic language or behavior, but yes all women and yes all men live in a society permeated by overt and subtle forms of sexism; forms inescapable by all women.

After the events in Ferguson, social and traditional media were awash with similar forms of denial. For instance, as Jon Stewart pointed out on his satirical news program *The Daily Show*, a Fox anchorwoman asserted, “this mantra of the unarmed black teenager shot by a white cop, that description in and of itself actually colors the way we look at this story.”[41] Here, the very facts of the case are said to be problematic in that they will “color” how the public feels about Darren Wilson, the police offer who shot Michael Brown. This is a form of denying reality and of denying the racism involved in this shooting. Hence, the response, “Hands up, don’t shoot,” made a similar statement to that of #YesAllWomen; it said perhaps not all cops are murderers, but yes, all black youth are and/or should be scared that they will be shot. This sentiment was further reflected in #IGottheTalk, which names how people of color are taught to handle being confronted by racial profiling and police harassment. Further, as Carol Anderson, associate professor of African American studies and history at Emory University, noted in a recent opinion piece in the Washington Post online, the media focus on black rage among the protestors in Ferguson problematically ignores the role of white rage:

> But the real rage smolders in meetings where officials redraw precincts to dilute African American voting strength or seek to slash the government payrolls that have long served as sources of black employment. It goes virtually unnoticed, however, because white rage doesn’t have to take to the streets and face rubber bullets to be heard. Instead, white rage carries an aura of respectability and has access to the courts, police, legislatures and governors, who cast its efforts as noble, though they are actually driven by the most ignoble motivations.[42]

To practice deep solidarity, to be willing to become both brain and shit, is to empathize with Hill’s black rage and to resist the white rage detailed by Anderson. It is to ask ourselves what it must feel like to have to have “the talk” with our children in a desperate attempt to save their lives from the very people our taxes pay to protect them. It is to never again rationalize that “but not all men . . . not all white people . . . not all rich people . . . not all Americans.”

There is a not insignificant affectual difference between these slogans and the "Yes we can," optimism of Barack Obama's 2008 presidential campaign. Feel the difference between, "yes, we can make a change," and "yes, we all have/will be under threat of violence and harassment" or "yes, we must put our hands up and even then we will have to ask to not be shot." The latter are depressing political statements, but it is through their very melancholic and lamenting character that I believe we can find hope for deep solidarity and for an authentic theology that resists the ideologies of globalization that are devouring us. When Hill sings, "Black rage is founded: who fed us self hatred / Lies and abuse while we waited and waited? / Spiritual treason, / this grid and its cages," "Victims of violence both psyche and body/Life out of context is living ungodly," and "Try if you must but you can't have my soul / Black rage is founded on ungodly control," she is singing the theology of the affect alien. She is saying with Ahmed that, "We can recognize not only that we are not the cause of the unhappiness that has been attributed to us but also the effects of being attributed as the cause."[43] The attempts at Hill's soul, the spiritual treason, and the violence to both body *and* psyche are not just effects of the original acts of oppression and abuse. It seems clear that for Hill, the threats to psyche and soul also stem from being forced to suppress such rage. "Life out of context is living ungodly," is a statement about taking the context of life as it has been materially (i.e., physically, psychically, and spiritually) shaped for black people, and denying its reality. To live godly, then, is to live out loud, to sing out one's black rage. Hence, it is in the naming of the sources of black rage where Hill finds she "doesn't fear so bad." Fearing and feeling bad remain when the rage is suppressed or its legitimacy denied. To be sure, when the rage is released happiness may not come, but that seems barely the point. That fear subsides and life can come into context can mean that life returns to godliness. As Ahmed has argued, at times leaving happiness for life "is to become alive to possibility."[44] This is the possibility that I find in the metaphor of "becoming brain/becoming shit." It is the possibility that comes when we abandon our desire for mastery in order to find deep solidarity with those who have been discarded by the masters.

Composting with God

For those of us who have resisted getting into the shit, even for those of us who through our financial disenfranchisement have felt the foundations of our heightened positions fracture, and so have fallen from the brain to the gut, *are we really ready to ask ourselves what it would mean to give up on our social and affectual capital? Are we ready to give up our proximity to the objects of success and happiness?* According to Antonio Negri and Michael Hardt, the post-Fordist society is one of the affect economy.[45] This economy includes wage laborers like hotel room attendants and waiters, but also those of us who are scholars and religious leaders, those of us whose product is affect. While this shift in economy represents for Hardt and Negri the possibility of a new mode of politics and social relations based on collaboration of the "multitude," in which, "immaterial workers... become a new kind of combatant, cosmopolitan bricoleurs of resistance and cooperation... [and] are the ones who can throw the surplus of their knowledges and skills into the construction of a common struggle against imperial power,"[46] it also comes with great risks of alienation. According to Hardt and Negri:

> When our ideas and our affects, our emotions, are put to work, for instance, and when they thus become subject in a new way to the command of the boss, we often experience new and intense forms of violation or alienation... The production of ideas, knowledges, and affects, for example, does not merely create means by which society is formed and maintained; such immaterial labor also directly produces social relationships.[47]

A turn toward the power of affect for both political resistance and violation can further push us to ask how deep solidarity might feel and whether we will risk becoming alienated from objects of "success."

When the adjunct faculty at the New School University (including a majority of the faculty at the Parsons School of Design) was embarking on a unionizing campaign with the United Autoworkers, the university administration posted signs that read, "Are you an artist or an autoworker?" What if we were to ask the same of ourselves, "Are we academics or autoworkers?" "Are we scholars or servers?" We need to take the tenor of this question seriously if we are going to understand the deep solidarity for which Rieger and Kwok have called. Further, we need to acknowledge why such a slogan was so successful at Parsons.[48] In other words, we must ask what it is about being called a laborer that is so threatening to the "professionals." Ahmed's understanding of happiness as pedagogical is, here, instructive. We are taught that a "professional" career is an object of happiness, and so one with which we should strive to be associated. We progressive theologians may not feel ontologically superior to laborers of a different sort, but I think we do often feel affectually superior. For instance, in discussions of deep solidarity Rieger has drawn on the following aphorism from Bertolt Brecht: "If I would not be poor, you would not be rich."[49] What if we reworked this line such that a wage-laborer said to the professor or the minister, "If I would not be *considered* unhappy, you would not be *considered* happy"? Or to put it another way, "If I would not be unprofessional, you would not be professional." What might it mean for us to challenge our own sense of class in this way? Through a reorientation of affect, might we find deeper solidarity? Could we even ask whether God is an artist or an autoworker? Can God contain what we too easily see as contradictory? Can God be both professional and unprofessional, a failure and a success, unimaginably small and unfathomably large? While we progressives might readily welcome the move from God as King to God as servant, will we be comfortable with God as precariat as much as (if not more than) God as poet? What about God as microbe as much as minister? What about God as feces?

To be sure, there are those of us who have been becoming shit for too long, who are brains and need to be acknowledged as such; there are those of us whose eventiveness of thought needs to be lifted upward so that it might finally be recognized for its creative potential. Yet there are those of us who need to be pulled down. Many of us in the 99 percent have not yet become fully disposable, although we are moving that way as we are pushed through the belly of the beast that is neoliberal capitalism. Hence, many of us will need to learn to decompose along with the compost, such that we might believe as Bennett does, that, "encounters with lively matter can chasten [our] fantasies of human mastery, highlight the common materiality of all that is, expose a wider distribution of agency and reshape the self and its interests."[50] I suggest that this is what the addition of feces does for our ontology. A hybrid sense of "being, becoming brain/becoming shit" helps us to trouble raced, gendered, and sexed theological and political foci on mastery, (re)productivity, and success. "Becoming brain/becoming feces" might even reshape the agency and the interests of God.

Might God, who is often seen as the brain of all brains, that giver and taker-on of forms, need to decompose with us, to be found in the shit pile of history, on the side of the expelled? What might it mean to think of an expelled God—one no longer found in the proper name of God, but in the movements of decomposition and recomposition found in the cycles of material life? In continuing her review of *Glas*, Spivak notes:

> The paradigmatic instance of the family's nominating function is burial: the family member's body is not allowed to become part of the undifferentiated and productive body of the earth. In the *Phenomenology*, Hegel writes: "the family keeps away from the dead this dishonouring of him by the desires of unconscious organic agencies and by abstract elements, puts its own action in place of theirs, and weds the relative to the bosom of the earth, the elemental individuality that passes not away. Thereby the family makes the dead a member of a community which prevails over and holds under control the powers of the particular material elements and the lower living creatures, which sought to have their way with the dead and destroy him" [*The Phenomenology of Mind*, trans. J. B. Baillie, Harper Torchbooks Edition (New York: Harper Torchbooks, 1967), p. 474].[51]

In *The Phenomenology of Mind*, absolute religion is reflected in scenes of the absolute family and so is representative of Absolute Knowledge.[52] Derrida's Genet column, the fecal column, runs alongside and betrays this search for mastery over "unconscious" organic agencies and "the powers of the particular material elements and the lower living creatures." If to succumb to the powers of the other vibrant material actants with which we are entangled is to destroy or dishonor a man, then perhaps a New Materialist theology will need to destroy and dishonor the God of the family name, of an absolute religion, *and* of absolute knowledge. This dishonored God is one that becomes brain *and* shit, vibrates in the material powers of decomposition and recomposition; and occupies with us and feels the black rage. It is this God—the God of the compost pile—that lures us to resist the anti-earth ideologies *and* theologies that are devouring us. For this is a God that will risk being devoured in order to find deep solidarity with those of us too often gobbled up and disposed of as waste.

Notes

1. Clayton Crockett and Jeffrey W. Robbins, *Religion, Politics, and the Earth: The New Materialism* (New York: Palgrave Macmillan, 2012), 126.
2. Crockett and Robbins, *Religion, Politics, and the Earth*, 67.
3. Crockett and Robbins, *Religion, Politics, and the Earth*, 153.
4. Mel Y. Chen, *Animacies: Biopolitics, Racial Mattering and Queer Affect* (Durham, NC: Duke University Press, 2012).
5. Gayatri Chakravorty Spivak, "Glas-Piece: A Compte Rendu," *Diacritics* 7:3 (Autumn 1977): 30.
6. Spivak, "Glas-Piece," 29.
7. Jane Bennett, *Vibrant Matter: A Political Ecology of Things* (Durham, NC: Duke University Press, 2010), 42.
8. Bennett, *Vibrant Matter*, 50.
9. Bennett, *Vibrant Matter*, 107.
10. Chen, *Animacies*, 13.
11. Chen, *Animacies*, 127.
12. Chen, *Animacies*, 127.
13. Joerg Rieger and Kwok Pui-lan, *Occupy Religion: Theology of the Multitude* (Lanham, MD: Rowman & Littlefield, 2012), 68.
14. Rieger and Kwok, *Occupy Religion*, 18.
15. Just what comforts need giving up should be a matter of urgent political and theological discussions (discussions I hope this essay helps to

prompt). I have been asked whether my call to become feces is a call to give up certain comforts that may actually provide power to middle class people, such as tenure and a 401(k) plan. The only answer I have to give is "maybe." Should full-time faculty give up the comforts of tenure in order to be in solidarity with the adjunct on government assistance? Maybe. Should the doctoral adviser have to give up her speaker honorarium in order to help pay the way for indebted and underfunded graduate students to attend conferences? Maybe. Should full-time salaried faculty members of the American Academy of Religion who participate in the Class, Religion, and Theology group collectivize their incomes with the group's adjuncts, graduate students, and unemployed PhDs? Maybe. Would this redistribution, and the loss of comfort that comes with knowing you will be able to pay your rent, feed your child, attend professional meetings, go on vacation, or have time to publish bind us closer together? Maybe. Conversely, would the giving up of comfort, or rather the theft of comfort through its redistribution—in this way just further support the institutions that for too long have profited off our cheap labor? Maybe. The call to become feces is not a call for unnecessary suffering but, rather, an invitation to reconsider what is needed for deep solidarity. Through the concept of becoming feces, I seek to question from where will come the political and theological energy to advocate for the right to live—to create, to rest, to play, to make love, to mourn, to rejoice, to eat, to be sheltered, to care, and be cared for, to be free of violence, and to feel rage and sadness in the face of the physical and psychic oppressions that work to rob you of these rights—no matter the status of one's employment. Perhaps it can come through the surplus energy of those whom because they have been considered shit have remained in a position of what Fred Moten and Stefano Harney call the fugitive academic [Stefano Harney and Fred Moten, *The Undercommons: Fugitive Planning & Black Study* (Wivenhoe/New York/Port Watson: Minor Compositions, 2013)]. According to Moten and Harney, "Like the colonial police force recruited unwittingly from guerrilla neighborhoods, university labor may harbor refugees, fugitives, renegades, castaways. But there are good reasons for the university to be confident that such elements will be exposed or forced underground. Precautions have been taken, book lists have been drawn up, teaching observations conducted, invitations to contribute made. Yet against these precautions stands the immanence of transcendence, the necessary deregulation and the possibilities of criminality and fugitivity that labor upon labor requires. Maroon communities of composition teachers, mentorless graduate students, adjunct Marxist historians, out or queer management professors, state college ethnic studies departments, closed-down film programs, visa-expired Yemeni student newspaper editors, historically black college sociologists, and feminist engineers. And what will the university say of them? It will say they are unprofessional. This is not an arbitrary charge. It is the charge against the more than professional. How do those who exceed the profession, who exceed and by exceeding escape, how do those maroons problematize themselves, problematize the university, force the university to consider them a problem, a danger?" (Harney and Moten, *The Undercommons*, 30). The more than professional, the fugitive academics who publish on the wrong blogs, who say the wrong thing at the meeting, who teach "triggering" courses (despite the risk to course evaluations), who distribute free copies of the readings to students who no longer can afford to attend the school, who get marked as "unprofessional, uncollegial, impractical, and naïve" (Harney and Moten, *The Undercommons*, 28), are also those marked as waste, the shit of academia. Yet this waste is needed; it is our labor upon labor that allows for the critical academic, the tenured professor critiquing the institution from which he receives comfort, to be legitimated (Harney and Moten, *The Undercommons*, 39). Yet transcendence is also unleashed by the waste—by the animacy found in the fugitive community. This immanent transcendence is nurtured anywhere that we are willing to be a problem for the university—to threaten it and place it in danger. What kind of problems might be created, for instance, if tenured faculty having collectivized their income with their junior colleagues could no longer afford to take time to publish? Would the threat to the University be worth the discomfort? Maybe. Maybe we do need to give up middle class comforts (or steal them in order to collectivize them), not so as to suffer more, but rather to participate in the excess produced by the more than professional. The more than professional, the marooned, the undercommons, do not refuse the University, but rather, as Jack Halberstam explains in their introduction to *The Undercommons*, "cannot be satisfied with the recognition and acknowledgement generated by the very system that denies a) that anything was ever broken b) that we deserved to be the broken part; so we refuse to ask for recognition and instead we want to take apart, dismantle, tear down the structure that, right now, limits our ability to find each other, to see beyond it and to access the places that we know lie outside its walls. We cannot say what new structures will replace the ones we live with yet, because once we have torn shit down, we will inevitably see more and see differently and feel a new sense of wanting and being and becoming" (Harney and Moten, *The Undercommons*, 6). This new sense that comes from tearing shit down, might also be that which comes from becoming feces.

16. Joerg Rieger, *No Rising Tide: Theology, Economics, and the Future* (Minneapolis, MN: Fortress Press, 2009), viii.
17. Lauren Berlant, *Cruel Optimism* (Durham, NC: Duke University Press, 2011), 1.
18. Berlant, *Cruel Optimism*, 106.
19. Judith Halberstam, *The Queer Art of Failure* (Durham, NC: Duke University Press, 2011), 120.
20. Halberstam, *The Queer Art of Failure*, 121.
21. This is not to say that those historically considered failures (the Woman, the Queer, the Person of Color, the Poor) are not a part of Occupy but, rather, to mark the movement by those whom have been considered "successes" or "victors" to reclaim a public body made up of all the failures in resistance to the system of success.
22. Judith Butler, "An Address to the Occupy Wall Street General Assembly," New York, October 23, 2011.
23. Judith Butler, "New Thoughts on Solidarity" Plenary Address, American Academy of Religion Annual Meeting, San Francisco, CA, November 14, 2011.
24. Amy Bingham, "Tea Party Debate Audience Cheered Idea of Letting Uninsured Patients Die." Abcnews.com. http://abcnews.go.com/blogs/politics/2011/09/tea-party-debate-audience-cheered-idea-of-letting-uninsured-patients-die/ (accessed September 1, 2014).
25. Judith Butler, "New Thoughts on Solidarity."
26. James Cone, *The Cross and the Lynching Tree* (Maryknoll, NY: Orbis Books, 2011).
27. Chen, *Animacies*, 2, emphasis added.
28. Chen, *Animacies*, 183.
29. Chen, *Animacies*, 182.
30. Melissa Gregg and Gregory J. Seigworth, eds. *The Affect Theory Reader* (Durham and London: Duke University), xi.
31. Sarah Ahmed, *The Promise of Happiness* (Durham: Duke University Press, 2010), 20.
32. Ahmed, *The Promise of Happiness*, 12.

33. Ahmed, *The Promise of Happiness*, 102.
34. Ahmed, *The Promise of Happiness*, 195.
35. Ahmed, *The Promise of Happiness*, 220.
36. Shelly Rambo, *Spirit and Trauma: A Theology of Remaining* (Louisville: Westminster John Knox Press, 2010) 73.
37. Rambo, *Spirit and Trauma*, 79
38. Aura Bogado, "Listen: Ms. Lauryn Hill's 'Black Rage' Responds to Ferguson." colorlines.com, http://colorlines.com/archives/2014/08/listen_ms_lauryn_hills_black_rage_responds_to_ferguson.html (accessed August 29, 2014).
39. See Ann Cvetkovich, *Depression: a Public Feeling* (Durham: Duke University Press, 2012).
40. Cvetkovich, *Depression*, 3.
41. *The Daily Show with John Stewart*. Comedy Central. August 26, 2014.
42. Carol Andersen, "Ferguson isn't about black rage against cops. It's white rage against progress." WashingtonPost.com. http://www.washingtonpost.com/opinions/ferguson-wasnt-black-rage-against-copsit-was-white-rage-against-progress/2014/08/29/3055e3f4-2d75-11e4-bb9b-997ae96fad33_story.html (accessed September 1, 2014).
43. Ahmed, *The Promise of Happiness,* 88.
44. Ahmed, *The Promise of Happiness,* 78.
45. Michael Hardt and Antonio Negri, *Multitude: War and Democracy in the Age of Empire* (New York: Penguin Press, 2004), 66.
46. Hardt and Negri, *Multitude*, 51.
47. Hardt and Negri, *Multitude*, 66.
48. The adjunct faculty at the New School eventually won a union after a politically and emotionally fraught battle with the administration.
49. Joerg Rieger, *Religion, Theology, and Class* (New York: Palgrave Macmillan, 2013), 197.
50. Bennett, *Vibrant Matter*, 122.
51. Spivak, "Glas-Piece," 34–35.
52. Spivak, "Glas-Piece," 36.

Chapter 5

Rethinking the New Materialism for Religion and Theology

Why Movements Matter Most

Joerg Rieger

The matters of religion and religious experience are made up of multiple components. In the study of religion, and even in some theological approaches, it is increasingly becoming clear that religious experience is never merely a matter of ideas and doctrines, nor is it merely a matter of disembodied individual experiences or mindless practices. Crude idealism, which focuses on abstract ideas, is insufficient for the study of religion, if not misguided. The same can be said of crude materialism, which focuses on matter in a deterministic way, although religion and theology have less frequently been studied from this perspective.

Nevertheless, these approaches still have plenty of followers. At one extreme are those who seek to explain religion and religious experience in terms of certain versions of the history of ideas. It is not only graduate students of religious studies who talk as if modern religion, for instance, and indeed modernity itself, had been invented singlehandedly by the ideas of philosophers like Immanuel Kant or theologians like Friedrich Schleiermacher. On the whole, variations of idealism are still prominent not only in the study of theology, where anti-materialism has found one of its homes, but also in various approaches to religion and religious experience.

At the other extreme are those who seek to explain religion and religious experience in crude materialist terms. In newly emerging fields like neurotheology, for instance, religion and religious experience are explained in terms of the function of the temporal lobes of the brain.[1] Crude political or economic materialism, seeking to explain religion and religious experiences in terms of political or economic structures, also have a few remaining adherents, but those approaches are mostly conjured up as ghosts of the past by anti-materialists who are keen to deny any impact of politics or economics on religion and religious experience.

In this context, new materialist approaches broadly conceived are helpful because they reclaim elements of the long-suppressed dialectical legacy of certain forms of materialism (and in some cases certain kinds of Hegelian idealism) for the study of religion and religious experience. At stake is not the abortive discussion of whether material or ideal factors are all-determinative, but how these factors influence and shape each other. While such approaches seem reasonable enough—even commonsensical—it is surprising that they still face substantial opposition, particularly in the realm of religious studies and theology. The so-called new materialisms are not the only approaches that reclaim the dialectic, but they are original proposals that deserve a closer look at a time when dialectics has faded from view in much of the study of religion and theology.

The way mental illness is being discussed helps exemplify the various positions. For the purposes of the following discussion, depression may serve as our example. Crude materialists argue that depression is best understood as chemical imbalances of the brain. As a result, it is treated with medications that help regulate these imbalances. The multi-billion dollar pharmaceutical industry profits from the prominence of this approach. Idealists, on the other hand, identify depression as located in the realm of ideas. For them, the treatment has to do with therapies that help depressed people think differently. The so-called power of positive thinking that is promoted by certain therapists as well as by a myriad of self-help books exemplifies a set of approaches that have proven to be quite lucrative as well. Each of these approaches has some benefits. Medication has done a lot of good for people suffering from depression, and positive thinking has saved others from self-destructive behavior.[2] Nevertheless, something is still missing here.

Better results are often achieved when there are overlaps between the various approaches. If they can afford it, people suffering from depression are often treated with both medication and the tools of psychological counseling. Moreover, the latter is not always based on pure idealism. A psychoanalyst, for instance, tries to figure out some of the material causes of depression. A behavioral therapist, to give another example, pays close attention to the effects of simple verbal instructions. Nevertheless, from a dialectical materialist position, a deeper analysis of the relation of material and ideal factors is required.

In order to get to the core of the problem of depression and to understand it from a dialectical materialist perspective, it is not sufficient to analyze what ideas are at work in a person's mind or what chemical imbalances are at work in the brain, although these things are important. Depression needs to be understood in terms of the intersection of mind, chemistry, and outside factors, reflecting on how these things mutually influence each other. The new materialisms in particular would insist that we cannot expect to find a simple determinism either way. Yet these reflections, which exemplify some of the concerns addressed by new materialists, are still too narrow because they often fail to consider broader structures and flows of power.

A very promising new approach to depression, developed from the perspective of pastoral counseling, exemplifies the broader perspective that is required. Here, depression is analyzed in terms of its occurrence under the conditions of neoliberal capitalism. This means that depression needs to be understood not only in terms of personal and social developments—both ideal and material—but also in terms of broader flows of power in a capitalist world. Large-scale economic factors, for instance, make a difference not only in politics and finance but also in the world of culture, religion, and ideas, as well as at the emotional level of individuals. Emphasizing these connections does not need to imply determinism. This kind of analysis, which exemplifies some of the implications of new materialist thinking even if it has not adopted that name, leads to different conclusions and to a different kind of therapy. As Bruce Rogers-Vaughn, a pioneer of this approach, has pointed out, depression is not only linked to capitalism, it can also help subvert it.[3] In other words, the concern of approaches that take material reality more seriously in dialectical fashion is never merely to analyze, but also to change, conditions. This is what Marx realized early on, in his famous Theses on Feuerbach.[4]

Based on these observations, I seek to explore three questions in this chapter: What might be the roadblocks that prevent materialist approaches from taking a hold in the study of religion and religious experience? What is the contribution of the new materialisms in this context? And where might we need to go beyond the discussions presented by the new materialists?

First, What Might Be the Roadblocks for Materialist Approaches?

It is both the benefit and the burden of materialist approaches that they introduce a certain complexity or even messiness into our understanding of religion and religious experience. Complexity can be a burden because it implies more work for the scholar and, at times, a certain inconclusiveness that some people may find frustrating. By comparison, definitions of religion and religious experience in terms of ideas tend to be relatively succinct and orderly. In addition, defining religion and religious experience in terms of ideas allows for relatively clear-cut identifications of certain kinds of religious orthodoxy, which seek to determine what (or who) is legitimate and what (or who) is not. In the still predominantly conservative cultural and political climate in the United States, such orthodoxy continues to enjoy high currency, although the situation seems to be changing slowly. However, the world of ideas also tends to be preferred in more liberal systematic theologies, as it allows greater orderliness and control over the subject matter.[5]

It is interesting to note that crude and deterministic materialisms tend to display characteristics similar to idealism, in that they appear to be succinct and orderly. If religious experience can be explained in terms of certain functions of the brain, for instance, the complexity and the messiness that is characteristic of religious experiences can be held at bay. Likewise, if religious experience is merely a function of politics or economics, there is no need to spend much time contemplating the complex realities of religion; one can then either reject it or embrace it for what it is. The attractiveness of non-dialectical methods endures for good reasons.

Perhaps the most misguided roadblock for materialism, which finds expression in popular discourse but is not "off-limits" in academic discourse, is the common complaint that people are "too materialistic." This appears to be a valid critique of life under the conditions of capitalism, but it is ultimately just another version of idealism, as the implication is that people are unduly driven by certain material concerns. Desires for a bigger house or for a new car and more money seem to come at the expense of spiritual concerns like ideals or beliefs. Here, whatever is considered to be material is bad and whatever is considered to be spiritual is good, and there is a simple choice that can be made by individuals as to whether they want to care about material or spiritual things. Several issues are overlooked: What if the desire for a new car is not merely a desire for material but also for spiritual matters, in that it promises safety, love, and comfort? The advertising industry has long figured this out. Moreover, what if religion is not the antithesis to what is material in general but provides a different way of being material and engaging material reality? In the Abrahamic religious traditions, for instance, there are strands that embody precisely this insight, teaching us that the problem is not material reality as such, but the kind of material reality that is embraced: is it that of the prophets and of Jesus or that of the successive empires of history?

To be sure, in many of the more recent approaches to the study religion complexity is acknowledged to some degree when scholars deal with multiple layers of ideas and texts or when scholars come to think of religion in terms of material embodiments, practices, or historical and cultural dynamics. This is a welcome development. Even the various postmodern turns to language and to culture have integrated some understanding of the material qualities of linguistic and cultural phenomena, informed by thinkers like Michel Foucault and his notion of discourse, which combines both ideal and material components.

What is still missing for the most part, however, are the kinds of complexities that include the broader structural power dynamics that have been the subject of dialectical materialisms since Karl Marx. The study of material embodiments and practices is often engaged without wondering where the tensions are and what drives these things. Moreover, Foucaultian discourse analysis is frequently performed without taking a look at who benefits from the flows of power and who does not. In some cases, even class identities are studied without an analysis of power, for instance when scholars investigate practices of working class religion without considering the influences of the ruling class and of capitalism.[6]

The benefit of overcoming the various roadblocks mentioned above is that dialectical materialisms are better equipped to take seriously the complexity of religion and religious experience, as they are receptive to a range of both ideal and material forces in the formation of religious experience. In this way, they are addressing the many criticisms made of the "tunnel-vision" of crude materialisms.

Nevertheless, this is not the end of our problems, as the challenge is bigger yet. Additional roadblocks emerge when any of these materialisms raise the question of power in the formation of religion and religious experience. Naming the dominant powers invariably leads to tensions and pushback, as these powers usually prefer to operate unconsciously and resist being analyzed for good reasons. If corporate advertising, for instance, is seen not only as spreading information or enticement to buy certain products but also as shaping people's most intimate sensitivities and preferences, new questions emerge that raise some concern. Those kinds of questions are not commonly addressed when religious experience is discussed. In my own work, I have encountered some of the fiercest resistance whenever I argued that religious experiences are shaped by various influences and forces, including monetary and political interests.

The benefit of dialectical materialisms is that they remind us that religious experience is not only complex and shaped by various ideal and material influences; it is also shaped by particular interests and powers whose effectiveness increases the more they operate underground and out of sight. This is what both crude materialists and idealists are missing. The most interesting materialisms insist on naming these powers, and this is where the resistance against materialism is most pronounced. In most (if not all) historical periods, the dominant powers that shape religious experience seek to ensure that they are not found out, pushing back against any approach that would question the pristine nature of religious experience.

For the Romans in the imperial period, for instance, it was more advantageous to believe in the gods sustaining the empire than in the empire sustaining belief in the gods. As a result, questioning these gods by pointing out their constructed nature amounted to treason. The early Christians, who raised such questions and talked openly about idolatry, like the apostle Paul,[7] were persecuted and killed. In capitalism, it is more advantageous to believe in an invisible hand sustaining the neoliberal free market than in the neoliberal market sustaining the belief in an invisible hand, and thus, those who raise questions about the genuine nature of this belief are quickly ostracized or simply ignored. To be sure, being ignored may be preferable to being persecuted and killed, but the end result is the erasure of opposition and thus the phenomena are not completely dissimilar.

Second, What Is the Contribution of the New Materialisms to the Study of Religion and Theology?

The new materialisms are of interest because they take up the dialectical heritage and thus help us broaden our investigations both in religious studies and in theology. They analyze how certain material factors (other than the ideal and crude material ones commonly studied) shape religious experience. This critical perspective leads to a constructive one that has often been overlooked, as religion can now also be examined in terms of the difference it makes to material reality. In addition, this perspective also allows for an awareness of the existence of alternative religious experiences that do not conform to the dominant powers and provides a deeper understanding of the nature and the promise of such experiences.

At the level of religious beliefs, philosopher William Connolly distinguishes between epistemic beliefs, which can be addressed and altered through reason and arguments, and beliefs that are linked to stronger identities "in which creed and affect mix together." When such beliefs are challenged, people experience "the tightening of the gut, coldness of the skin, contraction of the pupils, and hunching of the back." At the same time, these beliefs are also linked to feelings of abundance and joy.[8] Connolly, following Gilles Deleuze, calls this a "surplus," a term that complements my own earlier use of the notion of "theological surplus."[9] Such an understanding of religion allows for a two-way street between ideal and material realities.

In a recent book, religion scholars Clayton Crockett and Jeffrey Robbins reclaim Ludwig Feuerbach's materialist critique of religious experience with a positive twist: that human concerns play a major role in the formation of religious experience, they note, is not reason for rejecting religion but for reclaiming it.[10] Religious experience, in the account of the new materialists (via Karl Marx, G. W. F. Hegel, and Slavoj Žižek), is not merely false consciousness (keep in mind that no consciousness escapes falsehood altogether); it can become a force for empowerment and social change. Based on this insight, these new materialist scholars of religion invite us to take seriously material and physical realities and to reconceive the roles of economics, ecology, and energy in the production of religious experience.

This proposal merits closer attention, as it introduces new aspects into the study of religion and religious experience. In many conventional religious and theological discourses that are concerned with matters of social justice, for instance, religious experience is often discussed as affecting economics and ecology in straightforward fashion, but rarely as being affected by it and affecting it in dialectical fashion. Crockett and Robbins provide various examples for a dialectical relation of religious experience and material dynamics, keeping a close eye on broader material developments tied to ecology and the earth. In their work, the material perspective is broadened to include developments in the natural sciences, including the topic of energy. In a push beyond traditional physical concepts, Crockett and Robbins introduce new models of nuclear energy, with the conclusion that "the Earth [capitalized] becomes who it is through us if we have the energy and the foresight and courage to realize it."[11] Energy is thus both an external and an internal concept, as it passes between the natural sciences and the humanities, including the study of religion.

Such broadening of the dialectical materialist traditions is promising for various reasons. First, it includes a wealth of new insights produced in the natural sciences, from quantum physics to genetics and neurobiology. The natural sciences have come a long way from the days of Newtonian physics where cause and effect, subject and object, were easily distinguished and every question had a straightforward answer. This does not mean, however, that the natural sciences are now followed uncritically, as if their materialism would trump all other considerations. As the sciences are taken more seriously in the humanities and religious studies, we need to keep in mind that "sciences (and technologies) and their societies co-constitute each other," as Sandra Harding has pointed out from a feminist and postcolonial perspective.[12]

Second, new materialisms reshape and broaden our understanding of agency, as agency is not primarily rooted in terms of ideas and good intentions. Realizing a "mismatch between actions, intentions, and consequences," new materialists Diana Coole and Samantha Frost[13] advocate an open systems approach to the interactions between socioeconomic and environmental conditions, combining biological, physiological, and physical processes. It is not necessary, of course, to argue that actions and intentions do not matter at all, but scholars of religion and theology in particular can benefit from seeing them in a broader perspective. If the way to hell is paved with good intentions, as the old adage goes, we need to take another look at the problem of intentions, which is not merely a matter of good intentions that are not carried out. By the same token, matter needs to be considered as having agency in its own right, as new materialists emphasize "the productivity and resilience of matter."[14] These insights resonate with scientific thinking as developed in various fields, including quantum physics and molecular biology, resulting in a challenge of conventional ways of thinking about cause and effect. Matter, we are beginning to realize, is always in a process of becoming rather than merely being. One of the great advantages of these approaches is that the weight of having to produce transformation no longer rests on the shoulders of individuals alone.

Much of what we take for granted can be rethought from new materialist perspectives, even death itself. As Rosi Braidotti puts it, "death is overrated"[15] because our individual lives need to be seen in the larger context in which they affect and are affected by others, both human and nonhuman. This is comforting especially for progressives who, as Braidotti notes, are "just as human as others, only considerably more mortal."[16] Indeed, too many progressives have paid with their lives for the cause, from Jesus to Mahatma Gandhi and Martin Luther King. Braidotti is not a religious thinker, but her proposal to think "a Life [capitalized] that may not have 'me' or any 'human' at the center"[17] has profound parallels to various religious traditions, not the least of which is Christianity. In the words of the apostle Paul: "It is no longer I who live, but it is Christ who lives in me" (Gal. 2:20).

There is some potential here, although we need to gain further clarity about who the agents are in this model and how such agency can be realized under the conditions of capitalism, which seeks to harness every agency, human as well as nonhuman, for its own purposes. Crockett and Robbins, for instance, envision what they call a "postcapitalist world,"[18] but how do we deal with the capitalist world in which we live? We cannot just step outside of it. While it is commendable that in the new materialisms "the capitalist system is not understood in any narrowly economistic way but rather is treated as a detotalized totality that includes a multitude of interconnected phenomena and processes,"[19] we still need a clearer understanding of how capitalism works and how it shapes our reality as a whole, including religion. A start might be Coole and Frost's awareness of the "immense and immediate material hardship for real individuals" who lost their savings, their pensions, their houses, and their jobs in the meltdown of the economy after 2007.[20] Unfortunately the two authors fail to mention work and workers—one of the most crucial insights of dialectical materialism after Marx—and thus the exploitation of those who still have jobs and who, thus, have no choice but to function within the parameters of the neoliberal capitalist society.

Elsewhere in their book, Crockett and Robbins are clearer about capitalist reality, for instance, when they talk about "corporate control," which assures that even the immaterial (like digital reality) matters and takes on material force.[21] This is a crucial insight, because not even material reality itself escapes the flows of power in capitalism, a fact that is often overlooked. Nevertheless, these kinds of comments appear to be relatively infrequent in new materialist discussions and need to be deepened and expanded.

Third, Where Might We Need to Go beyond the New Materialisms?

In addition to the need to develop a stronger and more dialectical assessment of capitalist reality, many of the new materialisms are moving too fast when they look past some of the fundamental tensions traditionally addressed by dialectical materialism—namely, tensions that arise in the world of *labor*. It may come as a surprise, but this matter is highly relevant to the study of religion and religious experience, as we shall see soon.

In one collection of essays that is representative of the spectrum of thought in new materialism because it brings together significant representatives, there is hardly any mention of labor or work at all. When labor is mentioned, this is done mostly in order to note that the new materialisms need to go beyond the focus on labor that has been characteristic of materialisms in the past. Braidotti, who helped us see that even death can be relativized, develops her proposal as if labor were not even an issue, proposing instead a "biocentered egalitarianism" that "breaks the expectation of mutual reciprocity." She links this sort of egalitarianism to giving up ideas of retaliation and compensation.[22] Yet while retaliation and "tit-for-tat" may indeed not the most productive ways of relating to others and certainly not to the powers further up on the food chain, giving up notions of compensation and reciprocity altogether is not a real-life option for working people.

In order to go one step further, we need to go one step back. Where Crockett and Robbins "posit earth as subject"[23]—a factor that is indeed too often overlooked in religious discourse—what if we were to add working people who are doing most of the work that sustains humanity and its progress as subjects as well? This move does not have to exclude the earth, to be sure, but a dialectical materialism cannot afford to forget particular groups of people or, to use a term rarely heard in religious studies these days, certain *classes* of people that are not usually featured in religious discourse. In order to understand religion and religious experience more comprehensively, it needs to be taken into account that religion and religious experience shape up differently depending on who or what are their subjects, that is, who or what is involved in producing them. The problem is that, at present, not only are the contributions of the earth and of working people overlooked, they are systematically repressed. A broader understanding of religion and religious experience, and particularly of the various alternative forms, is not possible here.

It is no accident that among the most substantial roadblocks for materialist interpretations of religion in recent history have been efforts to downplay the role of oppressed people as authoritative sources of religious experience. This matches what is happening in the world of labor: workers' contributions are increasingly devalued by cutting salaries, benefits, and even hours at work—particularly for the lowest-paid workers, but also for many others. The workers of the Walmart corporation, the world's largest private employer, bear witness to this process.[24] If it is true that Walmart sets the pace for the rest of the economy (including white collar workers and professionals, if only by default), it is not inconceivable that Walmart's actions also have implications for the study of religion and what is considered appropriate. Large amounts of money certainly have implications for the practice of religion in communities where they are located. I suspect that some contemporary forces would rather acknowledge the earth as subject than oppressed people as subjects.

We can identify examples of efforts to downplay the role of oppressed people as authoritative sources of religious experience in the highly organized backlash against Latin American liberation theology, feminist theologies, and other theologies from the margins that keep demanding more than mere inclusion within the dominant system.[25] At the same time, the study of popular religion, which has become a hot topic, is not subject to similar backlashes, quite possibly because popular religious experience is usually studied in isolation from dominant religious experience and is therefore not perceived as threat. In other words, popular religion is not studied in terms of power and thus the question whether popular religious experiences might conflict with or challenge dominant religious experiences is rarely raised.

Here, remembering Marx's critique of Ludwig Feuerbach's materialism might be helpful. Going beyond Feuerbach, Marx observes that material objects and matter itself are not mere givens, but are produced by labor and commerce. Materialism needs to take into account the produced nature of matter, which means it needs to take into account not only the natural world but also the labor that is involved in producing matter. Matter, in other words, is never a static entity, as it is constantly produced and reproduced. This is true even for the natural world itself, a fact which did not escape Marx. Both nature and workers are constantly engaged in "changing the form of matter." Material wealth is always generated from the interplay of labor and nature.[26] While the new materialisms have done an excellent job deepening our understanding of the productive capacities of the "natural" world, far beyond what Marx and his contemporaries could have known, we must still provide a deeper analysis of the productive contributions of labor in the current economic situation. The dominant economic system is built on the exploitation of both nature and work, and one cannot be understood without the other.

For religious experience, this means that we need to study it in terms of material relations that are not universal—this is the mistake that many theists and atheists alike are making—but which are produced in the context of particular relations of power. Unless the materialist study of religious experience takes into account the history of how power is shaped and reshaped in particular social relations and in relation to social movements, paying attention to who benefits and who does not in a given system (including both people and earth), it will be unable to take seriously alternative forms of religious experience. That is precisely what was missing in Feuerbach, and that is what a contemporary materialist study of religion still needs to investigate.

A start is made by Jason Edwards in his conclusion to an important collection of essays on the new materialisms, when he argues for a return "to a kind of historical materialism that focuses on the reproduction of capitalist societies and the system of states, both in everyday practices of production and consumption and in the ideological and coercive power of states and the international system."[27] Edwards' focus on "material practices" is right on target, but this proposal could still benefit by addressing the realities of labor.

What difference would taking a closer look at labor and the reality of work make to new materialism and the study of religion and religious experience? First, nothing exists apart from production—production is a central fact of life, and we would not be able to exist without it. Taking a closer look at labor will help us to become more aware of the produced nature of all that surrounds us, religious experience included. This does not imply a negative judgment, as production is not a negative thing in itself, but it reminds us of the fact that nothing ever just "fell from heaven" and that there may be a few options involving alternative mechanisms of production.

Second, there is an odd sort of transcendence that occurs when produced objects are commodified. Marx uses the example of a table. In terms of its use value, there is nothing mysterious about a table. Wood, produced by nature, is altered by labor in order to produce a common thing to be used for particular purposes: a dining-room table, a desk, a kitchen table etc. A strange transcendence, however, enters in terms of the exchange value of the table. In economic exchanges, the table becomes a commodity and what matters is no longer the labor, the materials, or the use value but the profit that can be made when it is sold. Because profit is usually thought of as a relationship between things, what is concealed is that it is produced in a relationship between people.[28] In Marx's words, "A commodity is therefore a mysterious thing, simply because in it the social character of men's labour appears to them as an objective character stamped upon the product of that labour; because the relation of the producers to the sum total of their own labour is presented to them as a social relation, existing not between themselves, but between the products of their labour."[29] Marx compares this to religious ideas, where "the productions of the human brain appear as independent beings endowed with life."[30]

While the significance of labor and of human relationships is covered up in this process of commodification (this is what Marx means when he talk about "commodity fetishism"), it does not disappear. One of the new materialist thinkers, Sara Ahmed, makes a similar case.[31] Moreover, as Ahmed has argued, the example of the table also reminds us of other divisions of labor, manifest for instance in the division between who usually works at a desk and who usually works at a kitchen table. In this example, the kitchen table represents the racial and class-based divisions of labor, as work at the desk is supported by the domestic labor of black and working-class women.[32]

These reminders of labor and material relations can help us rethink our options when it comes to religion and religious experience. What Marx calls religion in the above example is how idealism views religion, as "independent things endowed with life"; but there is no reason why religion cannot also be viewed from a more materialist perspective that takes the relation of life to material realities into account. The same is true for the notion of transcendence: transcendence does not have to be an idealist notion. There are other ways to conceive of transcendence than as a dangerous abstraction from real life, for instance when it is defined not in opposition to immanence but as transcending one kind of immanence in favor of another.[33] While capitalism covers up the contributions of work, the study of religion can resist the cover up and thus benefit from taking work into account.[34]

Earlier, I welcomed the reference to material practices as crucial to any new materialism. There is nothing wrong with broadening this notion beyond activities that are immediately related to processes of production to include, as Edwards suggests, "all those practices involving material bodies—organic and nonorganic—that... can be seen as a totality of practices that reproduce the relations of production over time."[35] At a time when the nature of labor is shifting and more and more people are pushed into the informal sector where they might hold down casual jobs, so-called temp jobs, or no jobs at all, this is an important reminder. It also reminds us of the importance of the kind of work that is done without compensation, like housework or volunteer work, and other productive activities that people are doing off the clock. What Edwards calls "the constitution of experience through the manifold forms of material practice outside the immediate space of production,"[36] is valuable, especially when thinking about religious experience. This is what *Mujerista* theologian Ada María Isasi-Díaz has called "*lo cotidiano*," the "everyday" that is at the heart of life for most common people and for Latinas in particular.[37]

At the same time, in the current cultural climate in the United States there is very little reflection on how experience, religious experience included, is constituted through the regular processes of production at work, as labor is not a phenomenon that is commonly considered in the various fields of study, with few exceptions like the so-called "New Working Class Studies." Moreover, there has been hardly any work done on how class as a relational category shapes religion and religious experiences.[38] As a result, one of the most interesting questions yet to be investigated is how regular processes of production shape religious experiences of people. Without relating back to these very basic processes that occupy most of the waking hours of most people, despite the changes in labor just mentioned, we may not be able to develop a clear enough understanding of the importance of material practices more broadly conceived. Merely talking about the material practices of "everyday life" can also be misleading if it is not acknowledged that everyday life invariably is shaped by labor. Even leisure time is not "off the hook," because it is designed for the reproduction of our productive labor capacities and shaped by the trends and interests of the capitalist economy. That economy rests on the foundations of labor—that it constantly seeks to cover up.

The most important point is that the approach being suggested is not merely a matter of analyzing the impact of material practices on our lives, including religious experience. This approach is also a matter of identifying alternative ways of life and religious experiences that grow out of these material practices and the potential that reshaping processes of production might have for reshaping religious experience. To be sure, this is not a romantic dream about life far away from or above the dominant system. As Edwards correctly notes, "the material practices constitutive of modern life are the only grounds from which we could hope and expect to bring about important political and social transformations."[39] While material practices can make us compliant to the status quo, they also harbor the potential for resistance and for producing alternatives. The challenge for us is to find out which material practices are the most fertile ground for producing alternatives. In Marx's time, he identified industrial labor, which helped spawn the labor movement of his time. Today, that question is more complex, although I do not believe that it can be answered without the labor movement.[40]

Alternative religious experience is, therefore, closely linked to material practices and the alternative ways of life that grow out of them. Once again, this is where labor is important as it has produced powerful movements that have not only kept some of the worst abuses of capitalism in check but also built communities at the local, national, and international level that have made positive contributions to the world. As scholars of religion we need to investigate more thoroughly how alternative religious experiences shape up in the history of particular movements of exploited working people, in touch with the exploited earth, and what difference these experiences are making in the context of dominant religion. Neither scholars of religion nor theologians can produce such alternative religious experiences synthetically. They will have to study them while involving themselves in the resistance movements of our time, exploring what it all means for religion as a whole, including its practices, its doctrines and beliefs, and its ways of life. That was one of the concerns of the book *Occupy Religion*, co-authored with Kwok Pui-lan.[41] To bring it down to a formula: Materialism needs movements.

Conclusions

Some may wonder whether materialism ultimately threatens the work of scholars of religion and theology. This would be the case if religion and theology were understood as reflections on disembodied ideas and doctrines whose only connection to the world is that they need to be "put into practice." While that is indeed a widespread assumption, and the premise of many Sunday sermons, it is quickly becoming clearer that there are other options.

Dialectical thinking in the materialist tradition asks questions of power: What are we up against, where are the contradictions, what are the alternatives? Among the basic contradictions in capitalism are still the tensions between the ruling class and the working class. Knowing that we cannot escape these tensions—no matter how much we would prefer to hide behind scholarly objectivity—we need to decide what to do with them and how to rethink them in the present.[42] The question for politics and economics is clear: what side are we on, where do we hang our hat? The same question applies to matters of culture and religion. If we fail to answer these questions, the dominant status quo will answer them for us.

This awareness is surprisingly underdeveloped in the various discussions of the new materialism. For example, Crockett and Robbins write, "The political crisis we face today is that no one or nothing decides. The people have been rendered the object of market forces. The market... follows its own self-annihilating logic where only the few stand to benefit but none hold the power to control."[43] I am not so sure that is the case. The proverbial 1 percent is making decisions, and the rest of us had better realize this in order to organize and to exercise our options in the decision-making process. Unfortunately, class is rarely mentioned in these discussions. When it is referenced by Edwards, who is one of the few exceptions, it is only in order to note that class is not everything.[44] While this true, of course, one wonders how many voices are left today who would seriously argue that class is everything.

In this regard, it seems to me that the Occupy Wall Street movement has done more for us than any other movement in recent history. The proverbial tension between the 1 percent and the 99 percent, while not expressing a mathematically correct equation and while not claiming to be a full-fledged theory of class, may be more crucial than ever; not only for an understanding of politics and economics but also for an understanding of culture and, ultimately, of religious experience as well. As religion comes to life again in the discussion of new and dialectical materialisms, we grasp its alternative forms only when we grasp the larger structures of power that are at work in, around, and underneath neoliberal capitalism.

Notes

1. http://science.howstuffworks.com/life/inside-the-mind/human-brain/brain-religion.htm (accessed February 1, 2015).
2. Chemical dependencies, for instance, are common and medical drugs often merely cover up the deeper causes and problems of depression, providing a quick fix that does not resolve anything. Positive thinking can create a make-believe world that does not change reality and that misleads people into believing that they have power when they really do not. For a broader critique of positive thinking see Barbara Ehrenreich, *Brightsided: How the Relentless Promotion of Positive Thinking Has Undermined America* (New York: Metropolitan Books, 2009).
3. Bruce Rogers-Vaughn, "Blessed Are Those Who Mourn: Depression as Political Resistance," *Pastoral Psychology*, published online, November 6, 2013.
4. "The Philosophers Have Only Interpreted the World, in Various Ways; the Point is to Change It." https://www.marxists.org/archive/marx/works/1845/theses/theses.htm (accessed September 2, 2014).
5. In response to this problem, several theologians, myself included, have preferred to classify their work as constructive theology rather than systematic theology.
6. Sean McCloud, *Divine Hierarchies: Class in American Religion and Religious Studies* (Chapel Hill: University of North Carolina Press, 2007).

7. Romans 1:25: "They exchanged the truth for a lie and worshiped and served the creature rather than the Creator."
8. William E. Connolly, "Materialities of Experience," in *New Materialisms: Ontology, Agency, Politics*, ed. Diana Coole and Samantha Frost (Durham, NC: Duke University Press, 2010), 196.
9. Joerg Rieger, *Christ and Empire: From Paul to Postcolonial Times* (Minneapolis: Fortress Press, 2007).
10. Clayton Crockett and Jeffrey Robbins, *Religion, Politics, and the Earth: The New Materialism* (New York: Palgrave Macmillan, 2012).
11. Crockett and Robbins, *Religion, Politics, and the Earth*, 110.
12. Sandra Harding, "Beyond Postcolonial Theory: Two Undertheorized Perspectives on Science and Technology," in *The Postcolonial Science and Technology Studies Reader*, ed. Sandra Harding (Durham, NC: Duke University Press, 2011), 21.
13. Crockett and Robbins, *Religion, Politics, and the Earth*, 16.
14. Diana Coole and Samantha Frost, "Introducing the New Materialisms," in *New Materialisms: Ontology, Agency, Politics*, ed. Diana Coole and Samantha Frost (Durham, NC: Duke University Press, 2010), 7.
15. Rosi Braidotti, "The Politics of 'Life Itself' and New Ways of Dying," in *New Materialisms: Ontology, Agency, Politics*, ed. Diana Coole and Samantha Frost (Durham, NC: Duke University Press, 2010), 212.
16. Braidotti, "The Politics of 'Life Itself,'" 209.
17. Braidotti, "The Politics of 'Life Itself,'" 210, talks about "a profound love for Life as a cosmic force and the desire to depersonalize subjective life-and-death. This is just one life, not *my* life. The life in 'me' does not answer to my name: 'I' is just passing."
18. Crockett and Robbins, *Religion, Politics, and the Earth*, 145.
19. Coole and Frost, "Introducing the New Materialisms," 29.
20. Coole and Frost, "Introducing the New Materialisms," 31. I talk about this in terms of the "logic of downturn." See Joerg Rieger, *No Rising Tide: Theology, Economics, and the Future* (Minneapolis: Fortress Press, 2009).
21. Crockett and Robbins, *Religion, Politics, and the Earth*, 16.
22. Braidotti, "The Politics of 'Life Itself,'" 214.
23. Crockett and Robbins, *Religion, Politics, and the Earth*, xx.
24. See, for instance, the testimonies of Walmart workers who are organizing through OUR Walmart: http://forrespect.org/category/associate-voices/why-i-joined-our-walmart/ (accessed September 2, 2014).

25. By backlash, I do not mean legitimate debate, which is normal and healthy, but a wall of opposition that included ridicule, defamation, and other methods to discredit valid arguments. This organized opposition is not limited to the world of scholarship. In the 1980s, the Reagan administration, in its Santa Fe Document, identified Latin American liberation theology as a phenomenon that had to be held in check. See also the work of the so-called "Institute for Religion and Democracy."
26. "[Humans] can work only as Nature does, that is by changing the form of matter. Nay more, in this work of changing the form he [or she] is constantly helped by natural forces." Karl Marx, *Capital: A Critique of Political Economy*, vol. 1, book 1, trans. Samuel Moore and Edward Aveling; ed. Frederick Engels (Moscow: Progress Publishers, First English Edition of 1887), https://www.marxists.org/archive/marx/works/download/pdf/Capital-Volume-I.pdf (accessed September 2, 2012).
27. Jason Edwards, "The Materialism of New Materialism," in *New Materialisms: Ontology, Agency, Politics*, ed. Diana Coole and Samantha Frost (Durham, NC: Duke University Press, 2010), 283.
28. Marx talks about these issues in *Capital*, 46–47, but since he does not mention the terms exchange value and profit in this section, it is difficult to follow.
29. Marx, *Capital*, 46–47.
30. Marx, *Capital*, 47.
31. Sara Ahmed, "Orientations Matter," in *New Materialisms: Ontology, Agency, Politics*, ed. Diana Coole and Samantha Frost (Durham, NC: Duke University Press, 2010), 243.
32. Ahmed, "Orientations Matter," 248–254.
33. See our comments on transcendence in Joerg Rieger and Kwok Pui-Lan, *Occupy Religion: Theology of the Multitude, Religion in the Modern World* (Lanham, MD: Rowman & Littlefield, 2012), 71–76.
34. One of the fundamental shifts that I am suggesting in my book *No Rising Tide* is the move from a focus on redistribution to production.
35. Edwards, "The Materialism of New Materialism," 283.
36. Edwards, "The Materialism of New Materialism," 288.
37. See, for instance, Ada María Isasi-Díaz, *Mujerista Theology: A Theology for the Twenty-First Century* (Maryknoll, NY: Orbis Books, 1996).
38. To my knowledge, the notion of class as relationship has only been picked up lately in religious studies. See *Religion, Theology, and Class: Fresh Conversations after Long Silence*, ed. Joerg Rieger, in *New*

Approaches to Religion and Power (New York: Palgrave Macmillan, 2013). Class, when discussed at all, is often described in terms of stratification rather than in terms of relationship.

39. Edwards, "The Materialism of New Materialism," 292.
40. See also the use of the term "multitude" in Rieger and Kwok, *Occupy Religion*.
41. Rieger and Kwok, *Occupy Religion*.
42. See *Religion, Theology, and Class*.
43. Crockett and Robbins, *Religion, Politics, and the Earth*, 38–39.
44. See, for instance, Edwards, "The Materialism of New Materialism," 296.

Response
On the Virtue and Variety of Movement(s)

Jeffrey W. Robbins

The New Materialism has a counter-intuitive relationship with religion. This is due to both the old materialist, overtly critical attitude toward religion, as well as the continuing religious and spiritual concerns with the misplaced materialistic values associated with consumer capitalism. For the former, religion has been seen as the prime instance of false consciousness. For the latter, materialism is seen as necessarily materialistic, and thus, set in opposition to an authentic spirituality. The New Materialism provides an opportunity for rethinking many of the assumptions associated with these related terms. Specifically, the New Materialism offers alternative understandings of matter, materiality, and thus materialism itself. It provides a way of thinking such that the presumed opposition between materialism and religion can be seen instead in terms of a resource for spiritual renewal and political activation. Put alternatively, my thesis will be that the New Materialism is an altar call of sorts, an invitation to a kind of conversation away from the self and toward the earth. It is a materialism predicated on a metaphysics that sees matter in terms of energy transformation. Matter is not inert; instead, it moves. This gives movement to materialism, and is what makes the New Materialism genuinely non-reductive.

In what follows, I will first seek to debunk the presumed opposition between the material and the spiritual before turning to the ways by which the New Materialism I have in mind speaks to and can be clarified in conversation with the contributions to this volume, which I'll discuss in a slightly different order than they appear. But to be clear, in opposing the presumed contradiction between the material and the spiritual, I do not wish to disavow the spiritual, economic, and ecological concerns animating the effort to affirm both the sense of value that extends beyond a crude economic calculation and the sense of interdependency and interrelationship that is the basis upon which the concern with materialism qua materialistic culture is built. On the contrary, my argument is that the material is always already spiritual, economic, and ecological, and that the New Materialism thus is a way of dissolving these and related dualisms. What is needed is not a spiritual or religious revolution against the material, but a revolution in materialistic thought that recognizes its innately spiritual and political capacities.

(1) *The material is always already spiritual*: It is common to distinguish the material from the spiritual. Indeed, while much as been made of the differences between Pope Francis and his predecessor Pope Benedict XVI, they share a common concern with what Benedict once labeled the "tyranny of materialism."[1] For Benedict, this tyranny was associated with other pernicious features of contemporary Western thought and culture—specifically, relativism and secularism. For Francis, the danger of materialism is that it "robs us of our humanity."[2] Both criticisms are symptomatic of a reading of materialism that sees it as a product of consumer capitalism. As such, they see contemporary society's investment in material things as both theologically and morally problematic. Theologically, materialism is a form of idolatry in that it accepts money as the ultimate arbiter of value. Morally, it is a departure from Jesus's solidarity with the poor and the outcasts. In this way, the distinction between the material and the spiritual for both Francis and Benedict is one part of a more targeted comprehensive critique of the excesses of capitalism.

A less political form of this same sentiment can be seen in the teaching of the Dalai Lama. As he explains in his book, *Ethics for the New Millennium*, materialism is the underlying cause of the problems ailing modern society.[3] He sees materialism as indicative of a misplaced value that fundamentally misapprehends the true nature of reality. The materialism he has in mind reduces the human subject to that of a selfish and craving being whose wants and desires can be satisfied individually through the accumulation of things, often at the expense of others' wellbeing. In this way, materialism defies the truth of our profound interconnection and interdependence, and as such, can only satisfy our most superficial desires and only provide a transient and isolated form of happiness. What we really want—whether we know it or not—is a lasting contentment that can only come by way of the spiritual realization of ourselves as interdependent beings. In other words, if materialism is the underlying cause of the problems ailing modern society, then nothing less than a spiritual revolution is needed—a turning away from the material in a kind of transvaluation of values. As the Dalai Lama sees it, this fundamental spiritual truth is what all religions share in common.

These are important religious arguments that help to move people from their complacency with regard to the near total triumph of global capitalism. However, if the argument is that materialism only attends to human needs and wants at the most superficial levels, might there be an alternative way of conceiving of the material in such a fashion that it does not pit the material against the spiritual, or the self against others? Does the religious teaching, which affirms an intrinsic dignity of every human person that goes beyond any crude economic calculation, go deep enough? What if there is another way of conceiving of materialism such that there is no confusing or conflating it with the consumer culture of late capitalism?

I believe this is the great merit of the so-called New Materialism—namely, it is a non-reductive way of conceiving of the material such that it dissolves the distinction not only between matter and spirit, but also between nature and culture, brain and mind, body and soul, and religion and spirituality. The New Materialism I mean to affirm is not anthro-centric, but eco-centric.[4] It calls for the de-privileging of the human subject as the arbiter of truth and reality, and the re-situating of the human within the larger subject of the earth. Therefore, if the religious critique of materialism is that it effectively is a philosophy of self-gratification that amounts to an idolatry of the self, then the New Materialism is immune from that criticism because it calls for a conversion away from the self and toward the earth. In this way, we might say that the New Materialism is genuinely and intrinsically spiritual. Not in such a way that it pits the material against the spiritual, but by learning from the earth. We might learn from the earth the value of our flesh as embodied, material subjects. We might also learn to become more deeply spiritual by grounding our spirituality in the rhythms of the earth. This would not be an otherworldly spirituality, but one that seeks to repeat what the earth is already doing, and in so doing, contributing to an earth-driven political theology grounded in a proper sense of our common vulnerability and shared interdependency.

Likewise, it is common to associate the material with the inert. Correlatively, if material is thought to be inert, then the common association with materialism is that of a rigid determinism. For instance, consider the frequently articulated critique of Marx—namely, his thinking was scientifically reductive and economically deterministic. His materialism accomplished a dialectical reversal of Hegel, to be sure, but it was no less teleological, no less deterministic, and, which follows, no less responsible for the totalitarian form of politics that followed in its wake. The point of this critique is that by some Marxist materialist reckonings, human beings are nothing but products of economic forces beyond their control. Where the uninitiated see freedom, creativity, and hope, the Marx presented in this critique sees ideology producing a reified false consciousness in the form of an opiate for the masses. It is a steel-trap theory, where release is only made possible by the coming revolution—an inevitable implosion of social and political forces that never comes.

To move beyond this particular materialist impasse that confuses economic relations with matter that is static, we must rethink the metaphysics of materialism. That is to say, we must rethink materiality itself. The concept of plasticity provides me with the key to this effort. Specifically, the understanding of the plastic nature of materiality shows the movement, metamorphosis, and change of the very being of matter itself. Matter is not something that is static, but is always evolving. It is not merely given, but is auto-constituting. The philosopher Catherine Malabou refers to this as "ontological mutability," by which she argues that the essence of the matter of being is its own transformation.[5] The concept of plasticity denotes not only the reception of form, but also the giving of form; not a pure passivity, but activity—even agency, in the sense of epigenesis.[6] This is a non-reductive materialism because it indicates a basic freedom in becoming. Plasticity stands between freedom and determinism. The process of self-formation does not take place as if in a vacuum, but by a kind of rewiring of the brain, which Malabou likens to a rupture, or the annihilation of form. In other words, the becoming of being proceeds by way of resistance; it takes and remakes what is already given. The claim is that resistance is built into the very fabric of nature. As such, resistance is constitutive; it is not a reactive, but a creative and productive force. This is important not only because it demonstrates the never-ending movement of matter, but also because it provides a new materialistic basis for a contemporary politics of resistance.

(2) *Movement as Dance*: I consider much of my efforts in sketching a New Materialist theory of religion to be along the same lines as LaMothe's much more fully developed ecokinetic approach to the study of religion. She is correct that the New Materialism must not be construed as turning its back on the "old" materialists. The famed hermeneutics of suspicion—inaugurated by the likes of Feuerbach, Nietzsche, Marx, and Freud—is known for its respective critiques of religion, but it may be read differently, more affirmatively. For instance, a number of recent important works have reexamined Paul Ricoeur's so-called "masters of suspicion" in such a fashion that the *affirmative* quality of their religious thought has been emphasized.[7] The best known of these recent works is perhaps Van Harvey's *Feuerbach and the Interpretation of Religion* (1995).[8] Though Feuerbach himself was not included in Ricoeur's official pantheon, he has been considered by many to be a forerunner to the insights from Marx, Nietzsche, and Freud's critical attitude toward religion by his reading of theology as anthropology. For Harvey, however, just as important as Feuerbach's hermeneutic of suspicion is Feuerbach's *constructive* and evolving theory of religion, which has much significance beyond his well-known Hegelian critique of religion as alienation. Likewise Tyler Roberts' book on Nietzsche,

Contesting Spirit: Nietzsche, Affirmation, Religion (1998),[9] and James DiCenso's book on Freud, *The Other Freud: Religion, Culture and Psychoanalysis* (1999),[10] present challenges to the scholarly consensus that Nietzsche and Freud respectively are somehow enemies of religion. Roberts argues that Nietzsche's critique of religion is in fact an affirmative religious practice that eventuates not in the rejection or overcoming of religion, but rather its transformation and renewal. For DiCenso, though he admits that Freud's orientation toward religion is atheistic and generally critical, his work nevertheless yields constructive insights into the meaning of religion in culture.

I count LaMothe's work in this company. We see this especially in Run 3 of Part Two of her contribution to this volume, with her close reading of Feuerbach, Marx, and Nietzsche in terms of their valuation of "the ongoing rhythm of bodily becoming." Her point is an important one—namely, the dismissal of religion as false consciousness is less important than the kinetic images offered up that are life affirming, and that move beyond the somatophobic characteristic of so many idealistic theories of religion.[11]

However, from there she moves one step further, which is a leap I am unwilling to make. Specifically, her argument is that we must move *beyond matter* to movement. As I have tried to suggest earlier, the New Materialism as I understand it rejects this dichotomy between matter and movement. In this way, the allusions to "spirit"—which erupt not only in my work with Clayton Crockett, but also in the works of Rosi Braidotti and William Connolly that LaMothe examines in Part One—are not extrinsic but intrinsic, not a way of redeeming the material from being "merely material" but an immanent practice attentive to the movement of nature itself. In other words, I do not feel so compelled to move beyond matter to movement precisely because the New Materialism I have sought to help define is a matter already reconceptualized as movement.

Likewise, I am not so certain that the New Materialist project requires a philosophy of subjectivity, as LaMothe suggests when she asks, "Without a human subject, a materialist project itself is impossible: who will be there to care about the planet? Who will do the work involved in making a difference?" While I am certainly concerned about the survival of the human species, I am not persuaded that it is possible only by way of the identification of the materiality specific to the human body. LaMothe sees New Materialists like myself as "hamstrung by [their] professed commitment to dissolve the individual, anthropocentric subject." In contrast, I am suspicious of any definition of the human body that separates it from the earth. I too want to celebrate bodily movement and am grateful for the rich image of dance she elaborates for us here, put perhaps in the guise of an ecstatic whirling dervish, the notion of dancing to the rhythm of the earth's movement itself is what generates for me the visceral knowledge, the ecological awareness, and the intrinsic spirituality that I believe both LaMothe and I want.

(3) *Bowel Movements*: Throughout my work on the New Materialism, I have depended heavily on Freud. But to repeat: I read him affirmatively, not in order to dismiss or dispense with religion, but to identify its enduring appeal and its emancipatory potential. Now, challenged and inspired by Karen Bray's critical intervention from this volume, I want to expand on my reappraisal of Freud by taking even more creative license by offering up the following thesis: *excrement is the royal road to the unconscious.*

Bray draws on Affect Theory and Queer Theory to test the political claims of the New Materialism. She suggests the hybrid metaphor of becoming brain/becoming feces as an alternative to the nearly exclusive emphasis on the materiality of the brain for a discourse that takes embodiment seriously. This hybrid metaphor has a political saliency to it by attending to those historically considered disposable or who have traditionally been expelled from the social body. It also provides a more expansive conception of materiality in terms of Jane Bennett's "vibrant matter." That is to say, becoming feces involves production, consumption, and digestion, each marking the ongoing flux that is the continual process of energy transformation. It also indicates the nonhuman and inorganic agency key to theories of entanglement. This is a corrective I wholly affirm, and here want to consider the way that the bowel movement is yet another case in point of matter in motion.

To approach the topic, a question is in order: what is the matter with shit? I mean this as a serious question, and suggest two ways at answering it. First, asking what is the matter is another way of asking why something is a problem. When you ask someone "What is the matter?," you are not engaging in an ontological inquiry, but concerned with the person's emotional state. So one way of answering the question of why shit matters is because it stinks. But when and how did it come to stink? That is to say, at what point—and why—do we stop enjoying our excrement and begin to be repulsed by it? I know of no better way of introducing the fundamental seriousness of Freud's tripartite self than by pursuing this line of inquiry. A child is born into the world not yet knowing how to be disgusted. But eventually disgust is taken on like a second nature—interiorization coming at the price of alienation. In other words, the stink of shit is not an objective fact but a learned behavior or an acquired taste. What is the matter with shit, then, is that we come to think of ourselves as stinking bodies—or perhaps worse, we come to think of ourselves as separate and distinct from our stinking bodies.

Second, human waste is mistakenly thought of as dead matter—or, to return to the point made above, as inert. Such a view belies its movement. Human waste is produced and productive. Not only is it a source of food for the insect world, but it is the energy remains of the food already consumed, retaining as much as half of the energy that was sourced within the original food supply. In this sense, shit does essential work in decomposition. The better image than that of dead matter is to think of it as a battery pack to the ecosystem, a living transmitter that is teeming with life. In this case, what is the matter with shit is a material, ontological concern. Its matter is a metabolic ontology—a transfer and transformation of energy. From one perspective, waste is the mass of energy that is left behind after being put to its primary productive purpose. This is the anthropocentric subject perspective. But by thinking the earth as subject, waste is not secondary, but integral and immersive.

I do not mean to get too carried away with human waste's productive potential. Bray is correct that becoming feces is no less complex than becoming a brain, and it has the added benefit of getting us outside of our heads and into the world. But it is worth noting that this is a very different deprivileging of the brain than that of LaMothe's. For one, there is no dichotomy between matter and movement. There is no moving beyond materialism. On the contrary, the bowel movement proves a worthy subject matter for the New Materialism. Not only that, it rejects the appeal to transcendence, leaving us instead "feeling like shit." From Bennett's "vibrant matter" to Judith Butler's "precarious life," inherent to the becoming of feces is a radical vulnerability, of an unseen but still immanent agency at work. It is a redirection of political theology that does not rely on the concept of sovereignty. Nor does it merely suggest a form of political solidarity with the reject or the left behind that leaves the social hierarchy in place. Instead, the affirmation of fecal matter extends even—or perhaps especially—to its *nonproductivity*. There is loss, mourning, and grief, which may eventuate in a politics of refusal that is conceived and practiced on the basis of those who are thought *not to matter*.

The materiality of waste, then, is a new materialism of a certain kind of matter that does not matter, which is not to say that it is immaterial, spiritual, ethereal, or transcendent. On the contrary, this matter that does not matter is paradoxically the stuff of life. So I say it is the royal road to the unconscious because not only on the individual level is it the stinking reminder of that which we have disavowed and thereby unveils the process of disavowal itself, but also, on the collective level, it dispenses with any vestige of the utopic within the New Materialism and thereby unveils the mechanism of wish fulfillment still operative in even the most sophisticated theoretical discourses. This is what I take Bray to mean by her call for the reorientation of affect—not just learning about decomposition, but learning to decompose in order to be on the side of the expelled, from the exuberance of "Yes We Can!" to the despair, if not rage, of "Hands Up, Don't Shoot!"—both are genuine expressions that help to establish the conditions of possibility for political solidarity. It is the slash in the hybrid metaphor of becoming brain/becoming feces that prevents the New Materialism from making the mistake of keeping the dirty material carefully out of sight.

(4) *The Labor Movement*: Grateful does not do it justice; I am deeply indebted to Joerg Rieger. Not just for his contribution to this volume in which he poses the question of where we need to go beyond the New Materialism, but for the moral clarity of his theology. More than anyone, Rieger gets the point I have tried to make

regarding the religious value of the New Materialism in the terms of Marx's dialectical materialism—namely, the point is not merely to interpret reality, but to change it. This raises questions of power and ideology, and requires a naming of the economic and political influences that "operate under ground and out of sight." Rieger himself through the years has given various names to these influences, most significantly through the names and analysis of "Empire" and "Capital."[12] His work functions like that of a conscientious objector, occupying religious studies by his stubborn insistence that there must be a critical engagement with issues of class and must be a concern with matters of economic justice.

Confining myself to his chapter in this volume, I read Rieger in similar terms as I read Bray—specifically, as offering a sympathetic critique. Rieger's intervention is a reality check, a warning not to rush too quickly to a speculation of a postcapitalist world when, in the words of Coole and Frost, there is the "immense and immediate material hardship for real individuals" here and now.[13] Rieger is correct that any (old or new) materialism worth its salt must be concerned with work and workers. If not, we have turned our back on the systematic repression of workers. But most importantly—and in a way, splitting the difference between myself and LaMothe—is the connection Rieger makes between exploited working people and the exploited earth. So when coming to the question of which side we are on, or where we hang our hat, the choice between the poor and exploited of the earth and the earth itself is a false choice. What both require is a movement in terms of political organization.

To this, I have only one thing to add: It is regrettable that in the citation of the work of Malabou to develop our own conception of the New Materialism, Clayton Crockett and I neglected to emphasize the priority she places on the contemporary crisis of political subjectivity in terms of exploited labor.[14] It should be clear that for Malabou *neurobiological reality is just as much a biopolitical reality.* She claims a neuronal ideology that blinds us to the actual nature of brain plasticity, an ideology that stubbornly mistakes the brain as a machine or as a control center. This faulty understanding of the brain pairs up with an industrial mindset and managerial regime wherein the worker is the subject of control and domination. With shifts to a post-industrial economy in the West, however, this neuronal ideology also undergoes a shift—the brain as command-center or machine gives way to a faulty understanding of plasticity wherein plasticity is reduced to pure flexibility. In Malabou's words, "flexibility is plasticity without the genius."[15]

The genius here refers to the dual aspect of plasticity—that is, the concept of plasticity means not just the capacity to receive form or to be acted upon, but also, the capacity to give form or to resist. Therefore, the current neuronal ideology gets it only half right. Yes, the brain is not a fixed entity. It retains this permanent capacity to open up new neuronal pathways, to adjust to certain traumas, and to actualize almost an infinitude of possibilities. The brain demonstrates a remarkable resiliency and adaptability, adjusting and responding to almost any environment. Likewise the worker in a post-industrial economy: the worker is nomadic. To thrive in today's global economy, the worker must be willing and able to adjust to new market demands, adjust to the rapid pace of change, learn and relearn new skills and capacities. As such, it is a biopolitics of extreme flexibility that renders the subject almost entirely pliant, a blob of clay that can be molded and remolded at the market's whims.

But, this is only one half of plasticity. Malabou draws a hard line differentiating between flexibility and adaptability. What differentiates them is that the latter retains its resistance. Our current neuronal ideology blinds us to the possibility of resistance because such agency runs counter to the demands of a global workforce. We see in the brain only what we want to see. Or more precisely, the brain lacks self-awareness because of the biopolitical restraints imposed upon it by virtue of certain economic realities. This is the biopolitical culture of late capitalism as described by Fredric Jameson.[16]

If the plasticity of the brain is misconstrued only as flexibility, then this naturalizes and thus justifies all forms of the dislocation and exploitation of labor. The analysis here is akin to that offered by Gramsci in that the regime of control functions not by way of overt force or coercion—and this neuronal ideology is not a conspiracy orchestrated on high—but by the power of cultural hegemony. Also useful to understand here is the analysis of contemporary forms of alienation as detailed by the Italian theorist of post-industrial capitalism Franco Berardi wherein

> Productive life is overloaded with symbols that not only have an operational value, but also an affective, emotional, imperative or dissuasive one... The constant mobilization of attention is essential to the productive function: the energies engaged by the productive system are essentially creative, affective and communicational...
>
> Under these conditions *privacy* and its possibilities are abolished...
>
> Everywhere, attention is under siege.
>
> Not silence but uninterrupted noise, not Antoninioni's *red desert*, but a cognitive space overloaded with nervous incentives to act: this is the alienation of our times.[17]

This alienation effects a crisis in political subjectivity. The exploitation of physical labor has been superseded in some senses by the captivity of mental labor. And for Malabou, this captivity must be understood to be not only ideological, but material—because, after all, the mind is nothing other than the brain.

This is the contemporary crisis of political subjectivity, and what is desperately needed, according to Malabou, is a reschematized philosophy of freedom founded on a thoroughly materialized notion of resistance. Or as Berardi argues, the worker must "no longer be conceived as a passive object of alienation"—whether industrial or post-industrial, and whether physical or intellectual—"but instead as the active subject of a refusal."[18] In short, flexibility implies not only passivity, but also political docility, whereas adaptability suggests agency and the constitutive potential for resistance.

5) *(Energy) Conversion*: Lastly, I want to briefly comment on Kevin Minister's rehabilitation of conversion for Religious Studies and theology. I agree with his basic diagnosis of the limiting ways that the concept of conversion operates within much contemporary religious literature and practice. Specifically, within Evangelical Christian expressions, the fullness of the religious life has been collapsed into an individualized soteriology, forcing social justice and ecological concerns to the back burner. Such

notions of conversion reify the modern, secular norm that confines religion to an individual matter of private conscience. I also agree with the thesis that emerging analyses of political theology and the New Materialism offer an opportunity to reconceptualize conversion beyond an individual matter and in terms of an ongoing communal process.

Conclusion

In my work with Crockett, I have offered up a metaphysics of materialism based on energy transformation. Once again, this is not a static metaphysics or a crude, reductionist materialism, but one based on movement and change. Our radical proposal for energy does not feel the need to move beyond materialism, but rather, to move beyond the thermodynamic model. Its power lies in harnessing the energy waste in which the excess of heat is merely a nonproductive byproduct. By combining this with an appreciation of entropy, we are committed to finding new ways of producing energy, not by exploiting or subduing the earth's limited resources, but by *repeating what the earth is already doing.*

In this way, the New Materialism is an altar call of sorts, an invitation for a kind of conversion away from the self and toward the earth. By rooting itself in a conception of being as energy transformation, it is a materialism that moves. Even more, it is a materialism that recognizes and, hopefully, unleashes the energy already inherent in movements. As such, the theology of the New Materialism is a political theology—more precisely, a movement within political theology that embraces movements of various sorts, as the contributors to this volume have made abundantly clear. And finally, just as the New Materialism rejects the opposition between the material and the spiritual, it also breaks down the divide between transcendence and immanence. Or more precisely, it is a theology that places its faith not in an other-worldly transcendence, but in favor of an immanent transcendence that finds value in this vulnerable but, nevertheless, still infinitely creative world.

NOTES

1. See John Thavis, "In Africa, Pope Warns of Erosion of Values, 'Tyranny of Materialism'," in *Catholic News Service* (March 19, 2009), http://www.catholicnews.com/data/stories/cns/0901273.htm.
2. See Kerri Lenartowick, "Materialism Robs Us of Our Humanity, Warns Pope Francis," *Catholic News Agency* (September 29, 2013), http://www.catholicnewsagency.com/news/materialism-robs-us-of-our-humanity-warns-pope-francis/.
3. See Dalai Lama, *Ethics for the New Millennium* (New York: Riverhead Books, 1999), 14.
4. This has been developed most fully in the co-authored book with Clayton Crockett, *Religion, Politics and the Earth: The New Materialism* (New York: Palgrave Macmillan, 2012).
5. See especially, Catherine Malabou, *The Heidegger Change: On the Fantastic in Philosophy*, trans. Peter Skafish (Albany, NY: SUNY Press, 2011).
6. For instance, see Catherine Malabou, "The Future of Derrida: Time Between Epigenesis and Epigenetics," in *The Future of Continental Philosophy of Religion*, ed. Clayton Crockett, B. Keith Putt, and Jeffrey W. Robbins (Bloomington: Indiana University Press, 2014), 209–218.
7. For an elaboration on this observation, see Clayton Crockett and Jeffrey W. Robbins, *Religion, Politics and the Earth: The New Materialism* (New York: Palgrave Macmillan, 2012), 17–36.
8. Van Harvey, *Feuerbach and the Interpretation of Religion* (Cambridge: Cambridge University Press, 1995).
9. Tyler Roberts, *Contesting Spirit: Nietzsche, Affirmation, Religion* (Princeton, NJ: Princeton University Press, 1998).
10. James DiCenso, *The Other Freud: Religion, Culture and Psychoanalysis* (New York: Routledge, 1999).
11. For a genealogical analysis of how and why religion scholars have traditionally ignored the body, see Manuel A. Vasquez, *More than Belief: A Materialist Theory of Religion* (Oxford: Oxford University Press, 2011), especially 21–58.
12. See especially Joerg Rieger, *Christ and Empire: From Paul to Postcolonial Times* (Minneapolis, MN: Fortress Press, 2007); Joerg Rieger and Pui-lan Kwok, *Occupy Religion: Theology of the Multitude* (New York: Rowman & Littlefield, 2012).

13. Diana Coole and Samantha Frost, *New Materialisms: Ontology, Agency, and Politics* (Durham, NC: Duke University Press, 2010), 31.
14. See Catherine Malabou, *What Should We Do with Our Brains* (New York: Fordham University Press, 2008), especially 32–54.
15. Malabou, *What Should We Do with Our Brains*, 12.
16. Fredric Jameson, *Postmodernism, or, the Cultural Logic of Late Capitalism* (Durham, NC: Duke University Press, 1992).
17. Franco Berardi, *The Soul at Work: From Alienation to Autonomy*, trans. Francesca Cadel and Giuseppina Mecchia (New York: Semiotext(s), 2009), 107, 108.
18. Berardi, *The Soul at Work*, 23.

Conclusion

Between Religious Experience, Materialisms, and Social Movements

Edward Waggoner

This collection of essays is noteworthy in two respects. It is the first volume from religious studies scholars and theologians who critically appropriate tenets from "New Materialism" to argue constructive proposals about religious experience and socio-political movements. What LaMothe, Minister, Bray, and Rieger find congenial to their independent projects are "new materialist" claims about the vitality of matter, the universal distribution of agency, and the role of affect across multi-scaled "assemblages." This volume also signals new interest in academic work in religious studies that integrates descriptive and normative theory.

LaMothe, Minister, Bray, and Rieger challenge new materialists to refine some of their conceptual, analytical, and strategic moves. LaMothe contends that materialists miss a bodily basis for religious or spiritual ideas, values, and traditions. Minister argues that religious organizing—a form of "conversion"—is an overlooked engine for more equitable material conditions. Bray concludes that new materialist refrains such as "becoming brain" unwittingly erect barriers against more honest and effective solidarity among 99-percenters. Rieger warns new materialists not to underestimate the need for analytical categories of labor and class.

The single word, "Movement," is a signpost to this nexus of questions framed by religious experience and materialisms. Histories of materialisms, turns in religious practices and beliefs, kinesthetics of dance, expulsions of physical wastes and social beings, justice-oriented analytics and actions, and intellectual fads—all of these lie within the denotative and connotative range of "Movement." The term "Movement" could serve as a concise and affect-friendly reminder of the central questions of this volume, and also as a heuristic encouragement to new questions.

To explore "Movement" is one way to build vital and purposeful links between academy and culture now. Rigorous scholarship about religious experience and social movements, in connection with debates about contemporary materialisms, may awake latent synergies. The example of strategic alliances between diverse social movements and academic materialisms in the 1960s and 1970s offers some encouragement. While the lasting impact of those earlier conversations on social, political, and economic arrangements is a matter of contention, their ongoing influence in religious studies and theology is obvious.[1]

Methods, Risks, and Adjustments

Academic conversations about "Movement" may require participants to attenuate or suspend some methodological conventions of religious studies. Although feminist epistemologies and postmodern theories of culture have moved most fields in the humanities away from rigid divisions between fact and value, choices about methodology in religious studies can elicit intense and field-specific disagreements over whether or how a more porous relation between descriptive and normative theory should be negotiated.

There is a wide range of methodological positions in religious studies, and the differences between them are variously construed. General metaphysical commitments are features that can be useful in sorting out those differences. Such commitments are seldom expressed or discussed in religious studies. Roughly speaking, the field includes a large group of moderate realists, and relatively small pockets of neo-positivists and anti-realist social constructivists.

That is not an unusual distribution for fields in the humanities. However, in religious studies, these distinctions in general metaphysical commitments are complicated by disagreements over whether or in what sense "religion" exists as an object of study. These disagreements are also frequently inflected by, or conflated with, judgments about how to address the field's genealogical origins in Christian theology and escape its hegemonic influence across generations of scholarship.

Under these conditions, religious studies scholars, theologians, and proponents of contemporary materialisms who want to explore together the relation of religious experience and social movements need to be especially transparent about their working assumptions. What is meant by, "religious experience," or "religion?" What specific agendas for social, political, economic, and ecological change do participants have, and how are these agendas motivated? If discussants give concise answers to these questions at the outset, it will be far easier to perceive when, how, and why subsequent normative proposals overlap.

Creative and compelling scholarship at the intersection of religious studies, theology, and materialism may very well generate methodologies that can enliven scholarship that is more field-specific. Confidence for that outcome stems not from a vague, underwarranted belief that any project that is interdisciplinary will be widely transformative, but from the observation that to *begin* academic conversations with such explicitly normative goals is uncommon—especially, and for the reasons mentioned above, in religious studies.

If necessity is a goad to innovation, then scholars who participate in conversations about "Movement" should expect that their efforts will yield methodologies, ontologies, and ethical positions that are unlooked-for and surprising—perhaps even unsettling. If normative aims remain explicit, and intellect and affect acknowledged as co-constitutors of rigorous and truly open-ended discussion, then participants will have strong license to risk creative proposals that may not be given a careful hearing in their home disciplines.

Participants may realize that they must modify already-established methodologies in order to get going; or, they may turn immediately to the conceptual conundrums already visible in their normative goals. Either way, if co-participants help one another to reidentify periodically the methodologies, ontologies, and ethical priorities implied by subsequent versions of both the descriptive and normative aspects of their proposals, they will be able to trace salient shifts.

Scholars will be more likely to break new ground in theorizing religious experience and social movements if they let shifts in their working assumptions accumulate, and refine them through critical discussion. The process will certainly enhance the cogency of normative proposals. The authors of such proposals may also, as they move forward, arrive at generative descriptions or theories of religious experiences, claims about what "religion" is, and insights about what contemporary materialisms reveal or occlude. These results, and related methodologies and analytical strategies, could prove useful to home-discipline colleagues who work on other questions.

Strategies for Conversation

Conversations between religious studies scholars, theologians, and theorists of contemporary materialisms about imbrications of religious experience and social movements are only beginning. But the authors of this volume (and new materialists elsewhere) have already provided resources—in the form of concepts, themes, problems, theses, and questions—that can be carried forward to the next round of conversations. There are, no doubt, many viable approaches to maximizing the benefit of those resources. Here I will suggest four.

First, scholars can identify and discuss the explicit and implicit normative goals in the service of which crucial concepts, themes, problems, theses, and questions operate. The advantages to this approach are several: it carries forward narratives that help readers to interpret them; it encourages readers to notice their own affective responses to what they take the author's goals to be; it suggests a clear, initial question for immanent critique. How effectively are these resources marshaled for the normative goal? Importantly, by attending to normative goals directly, scholars reissue the invitation to craft normative proposals.

Second, scholars can move toward the next round of conversations about religious experience, social movements, and materialisms by articulating what epistemologies, ontologies, and metaphysics are implied by the arguments in their respective proposals. It is obvious why that would be useful: proposals for social, political, and economic change imply diverse agencies and causal relations. Epistemologies and ontologies already receive extensive, frequent, and explicit attention from new materialists. Although it would take more effort to articulate what is implicit in proposals from religious studies scholars and theologians, it would make critical comparisons possible.

Third, for the next round of conversation about religious experience and social movements, in the context of materialisms, scholars can benefit from extant work by focusing on specific concepts that recur across proposals. Observations about patterns of usage, and similarities and differences between the work that key concepts do in specific descriptions and arguments, can help participants to critique proposals. A focus on conceptual patterns will also help scholars discern which aspects of religious experience and social movements are receiving the most emphasis. That leads naturally to questions about why those aspects rather than others garner attention.

Fourth, scholars can benefit from work by LaMothe, Minister, Crockett and Reader, Bray, Rieger, Robbins, and others who are engaging new materialism by noticing what they do not say or ask, issues that would be important to address in the next conversations about religious experience and social movements. It is a truism that no proposal can be about everything. It seems also true that some questions are not asked because subject-positions occlude them. Some questions simply do not exist, unless and until certain other questions are uttered. A search for unasked questions can be an explicit component of the next conversations about "Movement."

Whether these or other approaches are adopted by scholars who participate in the next round of conversations, the cardinal point is this: progress toward normative theories that are creative and compelling will depend on processes both social and cumulative. It requires scholars to convene, in-person.[2] It also requires them to enact strategies for remembering, discussing, and transforming the specific concepts, themes, theses, and questions already offered.

To Weigh Proposals

In her introduction to this volume, Tamsin Jones identifies a question that arises—in one form or another—from the essays by LaMothe, Minister, Crockett and Reader, Bray, Rieger, and Robbins: how well do new materialist claims about human subjectivity support proposals for socio-political activism? This is an important, evaluative question that, in conversations about "Movement," should be applied to all claims about human subjectivity—whether new materialist or not.

To ask whether a proposal or viewpoint provides adequate support for socio-political activism could be to ask *Will individuals or groups who accept this narrative as a trustworthy guide to how the world is, and who they are in it, feel as a result more empowered to resist unjust socio-political arrangements and work actively to create better ones?* That is a question best answered after the fact, by asking specific individuals or groups whether or not they do—by accepting this or that narrative—feel themselves strengthened in the relevant respects.

To gloss the question in that way reminds participants in conversations about "Movement" how difficult it is to demonstrate causal links between theoretical proposals and changes to political, economic, or ecological processes. It is easy to forget or not actually believe that contexts influence theory making. But it is also easy for theorists in the humanities to elide unconsciously the distinction between a proposal's conceptual coherence and effective results.

It is possible to imagine several outcomes in which conceptual coherence and effective results are two independent, or only weakly dependent, variables. For example, there could be: a normative proposal (or theory) that features a conceptually incoherent account of what it means to be a human agent, but proves successful in mobilizing persons to join a social or political movement; or, a narrative that is effective in virtue of, rather than in spite of, its incoherency; or, a proposal that remains ineffective despite nearly perfect conceptual coherence.

Because part of what is at issue in such discussions are changes in material arrangements of bodies and things, it seems wise to make the evaluative criteria for adequacy more explicit. If participants are transparent about what sort of adequacies are intended, critical responses can mirror that choice or argue why other grounds for critique will be more generative for conversation.

If participants do not have a way to evaluate the extra-textual effectiveness of descriptions of human subjectivity, they can rely on well-known forms of immanent critique. One option is to evaluate the relation between various claims within a given proposal (e.g., are the claims about human subjectivity conceptually adequate to the proposal's vision for socio-political arrangements?). Another option is to evaluate the relation between the proposal's claims about human subjectivity and someone else's vision for socio-political arrangements (e.g., are the claims conceptually adequate to the reader's own vision?).

A Composite of Human Subjectivity

Scholars of "Movement" share a basic aim: to develop ways of thinking, acting, feeling, speaking that resist de-forming social, political, economic, and ecological arrangements, and support justice. One of the best ways to convey how LaMothe, Minister, Crockett and Reader, Bray, and Rieger have worked toward this basic aim is to present a composite of their descriptions of human subjectivity.

This human subjectivity is hard working. In part, we know this subjectivity as it is expressed in the concrete, material things that it makes, as well as in processes of production that exploit subjectivities. We learn of this subjectivity from its "turns"; and from the movements in which it turns; and from the way other material bodies move in response to its work. "Bodies under pressure" can create their own spaces and times by moving in directions of justice. Human subjectivity is a flow of energy.

This human subjectivity dances.[3] Everything the human dancer touches, sees, hears, tastes, or smells, gains value. It whirls ecokinetic images outward, and into which it dances; and dancing there, it feels the rhythms of the earth itself. The world changes because of this dancer. With every perception, every movement, humans alter themselves and their contexts. This is rhythm as a human being. This is a subjectivity that listens, above all, to joy.

This human subjectivity fashions itself with choices: a kind of affective transformation. This is subjectivity that wills to change, and changes through affective solidarity with those who are dishonored and expelled by dominant systems. Even God comes to be changed through affective solidarity. This is human subjectivity behind megaphones, in sweat. A subjectivity of desire and waste: alive within fecal forms.[4] Subjectivity is solidarity.

This is a subjectivity that moves within tight spaces. This is human subjectivity that lives along the edge of chaos.[5] It will not ever be quite what "matter" or modes of production or theories of history claim that it is. This is subjectivity for which relations with others are more important than any notion of immanence or transcendence. It follows the bodies, follows the money. This is a subjectivity where movements matter.[6]

NOTES

1. One could mention (among others): Simone de Beauvoir, Gilles Deleuze, Michel Foucault, Jacques Derrida, Luce Irigaray, and Jürgen Habermas.
2. They also need to remain in touch with social movements.

3. This is a term that Kimerer LaMothe uses. See her chapter in this volume.
4. This is a term that Karen Bray uses. See her chapter in this volume.
5. This is a phrase that Clayton Crockett and John Reader use. See their chapter in this volume.
6. This is a phrase that Joerg Rieger uses in the title of his chapter in this volume.

Contributors

Karen Bray is a PhD candidate in Theological and Philosophical Studies in Religion at Drew University. Her research employs queer and feminist approaches to affect, temporality, and work in order to construct a counter-capitalist political theology. For the last three years, she has taught undergraduate courses in Queer and Decolonial Theology, Sexual Ethics, New Materialism, World Christianities, and Affect Theory at The New School in New York City. She is a candidate for ordination with the Unitarian Universalist Association.

Clayton Crockett, Professor and Director of the Religious Studies Program at the University of Central Arkansas in Conway, Arkansas. He is the author of *Religion, Politics and the Earth: The New Materialism* (2012). He is also a Contributing Editor of the *Journal for Cultural and Religious Theory* and co-editor of the Columbia University Press book series "Insurrections: Critical Studies in Religion, Politics, and Culture."

Tamsin Jones is Assistant Professor of Religion at Trinity College in Hartford, Connecticut, where she teaches courses in the history of Christian thought, gender and religion, and the philosophy of religion. She is the author of *A Genealogy of Marion's Philosophy of Religion: Apparent Darkness* (2011) as well as articles in *Modern Theology* and *Journal of Religion*. She is currently working on the concept of "religious experience" as it is discussed in continental philosophy and against the backdrop of trauma theory.

Kimerer L. LaMothe is a dancer, philosopher, and scholar of religion who taught at Brown and Harvard Universities before moving to a farm in Upstate New York. She is a recipient of fellowships from the Radcliffe Institute of Advanced Study and the Center for the Study of World Religions and the award-winning author of five books, including *Why We Dance: A Philosophy of Bodily Becoming* (2015) and *Nietzsche's Dancers* (2006).

Kevin Minister is an Assistant Professor of Religion in the College of Arts and Sciences at Shenandoah University in Winchester, Virginia. His current book project explores the role of religious organizing in labor practices and workers' movements.

John Reader is Associate Research Fellow with the William Temple Foundation, University of Chester, a parish priest, theological educator, and practical theologian. His publications include *A Philosophy of Christian Materialism: Entangled Fidelities and the Public Good*, co-authored with Chris Baker and Tom James (2015), *Theological Reflection for Human Flourishing: Pastoral Practice and Public Theology*, co-authored with Helen Cameron, Victoria Slater, and Christopher Rowland (2012), *Christianity and the New Social Order*, co-authored with John Atherton and Christopher Baker (2011), and *Reconstructing Practical Theology: The Impact of Globalization* (2008).

Joerg Rieger is Wendland-Cook Professor of Constructive Theology at Perkins School of Theology, SMU. His most recent books include *Religion, Theology, and Class* (editor, 2013), *Occupy Religion: Theology of the Multitude* (with Kwok Pui-lan, 2012), *Grace under Pressure: Negotiating the Heart of the Methodist Traditions* (2011), *Globalization and Theology* (2010), *No Rising Tide: Theology, Economics, and the Future* (2009), and *Christ and Empire* (2007), with various translations into Portuguese, Spanish, German, and Chinese. Rieger is editor of the Palgrave Macmillan book series "New Approaches to Religion and Power," and co-editor of the Rowman and Littlefield book series "Religion in the Modern World."

Jeffrey W. Robbins is Chair and Professor of Religion and Philosophy at Lebanon Valley College, where he also serves as the director of the American Studies program and as the Faculty Mentor for the Allwein Scholars Program. He is a Fellow of the Westar Institute and is an affiliated faculty member of the Global Center for Advanced Studies. He is the author or editor of various books, including most recently *Radical Theology: A Theological Method for Change* (forthcoming) and *Radical Democracy and Political Theology* (2011), and with and with Clayton Crockett, *Religion, Politics and the Earth: The New Materialism* (2012), He is also a Contributing Editor of the *Journal for Cultural and Religious Theory* and co-editor of the Columbia University Press book series "Insurrections: Critical Studies in Religion, Politics, and Culture."

Edward Waggoner is Assistant Professor of Theology, in the Rt. Rev. Sam B. Hulsey Chair in Episcopal Studies, at Brite Divinity School, in Fort Worth, Texas, where he teaches constructive, systematic, and liberation theologies. His current project is a critique of religious support for militarization in the United States.

Index

9781137568434